# Interactive Mathematics Program

**I M P**

## Integrated High School Mathematics

YEAR **2**

## Solve It!

Dan Fendel and Diane Resek
with
Lynne Alper and Sherry Fraser

KEY CURRICULUM PRESS
Innovators in Mathematics Education

This material is based upon work supported
by the National Science Foundation
under award number ESI-9255262.
Any opinions, findings, and conclusions
or recommendations expressed in this
publication are those of the authors
and do not necessarily reflect the views
of the National Science Foundation.

Key Curriculum Press
P.O. Box 2304
Berkeley, California 94702
510-548-2304
editorial@keypress.com
http://www.keypress.com

10 9 8 7 6 5 4 3 2 1   01 00 99 98 97
ISBN 1-55953-258-0
Printed in the
United States of America

## Project Editor
Casey FitzSimons

## Editorial Assistant
Jeff Gammon

## Production Editor
Caroline Ayres

## Art Developer
Jason Luz

## Mathematics Review
Rick Marks, Ph.D., Sonoma State University
     Rohnert Park, California

## Teacher Reviews
Daniel R. Bennett, Kualapuu, Hawaii
Larry Biggers, San Antonio, Texas
Dave Calhoun, Fresno, California
Dwight Fuller, Clovis, California
Daniel S. Johnson, Campbell, California
Brent McClain, Hillsboro, Oregon
Amy C. Roszak, Roseburg, Oregon
Carmen C. Rubino, Lakewood, Colorado
Jean Stilwell, Minneapolis, Minnesota
Wendy Tokumine, Honolulu, Hawaii

## Multicultural Reviews
Mary Barnes, M.Sc., University of Melbourne,
     Cremorne, New South Wales, Australia
Edward D. Castillo, Ph.D., Sonoma State University,
     Rohnert Park, California
Joyla Gregory, B.A., College Preparatory School,
     Oakland, California
Genevieve Lau, Ph.D., Skyline College,
     San Bruno, California
Beatrice Lumpkin, M.S., Malcolm X College (retired),
     Chicago, Illinois
Arthur Ramirez, Ph.D., Sonoma State University,
     Rohnert Park, California

## Cover and Interior Design
Terry Lockman
Lumina Designworks

## Production Manager
Steve Rogers, Luis Shein

## Production Coordination
Diana Krevsky, Susan Parini

## Technical Graphics
Kristen Garneau, Natalie Hill, Greg Reeves

## Illustration
Tom Fowler, Evangelia Philippidis, Sara Swan,
Diane Varner, Martha Weston, April Goodman Willy

## Publisher
Steven Rasmussen

## Editorial Director
John Bergez

# *Acknowledgments*

Many people have contributed to the development of the IMP curriculum, including the hundreds of teachers and many thousands of students who used preliminary versions of the materials. Of course, there is no way to thank all of them individually, but the IMP directors want to give some special acknowledgments.

We want to give extraordinary thanks to the following people who played unique roles in the development of the curriculum.

- **Matt Bremer** did the initial revision of every unit after its pilot testing. Each unit of the curriculum also underwent extensive focus group reexamination after being taught for several years, and Matt did the rewrite of many units following the focus groups. He has read every word of everyone else's revisions as well and has contributed tremendous insight through his understanding of high school students and the high school classroom.

- **Mary Jo Cittadino** became a high school student once again during the piloting of the curriculum, attending class daily and doing all the class activities, homework, and POWs. Because of this experience, her contributions to focus groups had a unique perspective. This is a good place to thank her also for her contributions to IMP as Network Coordinator for California. In that capacity, she has visited many IMP classrooms and has answered thousands of questions from parents, teachers, and administrators.

- **Lori Green** left the classroom as a regular teacher after the 1989-90 school year and became a traveling resource for IMP classroom teachers. In that role, she has seen more classes using the curriculum than we can count. She has compiled many of the insights from her classroom observations in the *Teaching Handbook for the Interactive Mathematics Program*™.

- **Bill Finzer** was one of the original directors of IMP before going on to different pastures. Though he was not involved in the writing of Year 2, his ideas about curriculum are visible throughout the program.

- **Celia Stevenson** developed the charming and witty graphics that graced the prepublication versions of all the IMP units.

Several people played particular roles in the development of this unit, *Solve It!*:

- Matt Bremer, Janice Bussey, Donna Gaarder, Theresa Hernandez-Heinz, Linda Schroers, and Adrienne Yank helped us create the version of *Solve It!* that was pilot tested during 1990-91. They not only taught the unit in their classrooms that year, but also read and commented on early drafts, tested almost all the activities during workshops that preceded the teaching, and then came back after teaching the unit with insights that contributed to the initial revision.

- Melody Martinez, Greg Smith, and Cathie Thompson joined Matt Bremer for the focus group on *Solve It!* in November 1994. Their contributions built on several years of IMP teaching, including at least two years teaching this unit, and their work led to the development of the last field-test version of the unit.

- Dan Branham, Dave Calhoun, Steve Hansen, Gwennyth Trice, and Julie Walker field tested the post-focus-group version of *Solve It!* during 1995–96. Dave and Gwennyth met with us to share their experiences when the teaching of the unit was finished. Their feedback helped shape the final version that now appears.

In creating this program, we needed help in many areas other than writing curriculum and giving support to teachers.

The National Science Foundation (NSF) has been the primary sponsor of the Interactive Mathematics Program. We want to thank NSF for its ongoing support, and we especially want to extend our personal thanks to Dr. Margaret Cozzens, Director of NSF's Division of Elementary, Secondary, and Informal Education for her encouragement and her faith in our efforts.

We also want to acknowledge here the initial support for curriculum development from the California Postsecondary Education Commission and the San Francisco Foundation, and the major support for dissemination from the Noyce Foundation and the David and Lucile Packard Foundation.

Keeping all of our work going required the help of a first-rate office staff. This group of talented and hard-working individuals worked tirelessly on many tasks, such as sending out units, keeping the books balanced, helping us get our message out to the public, and handling communications with schools, teachers, and administrators. We greatly appreciate their dedication.

- Barbara Ford—Secretary

- Tony Gillies—Project Manager

- Marianne Smith—Communications Manager

- Linda Witnov—Outreach Coordinator

We want to thank Dr. Norman Webb of the Wisconsin Center for Education Research for his leadership in our evaluation program, and our Evaluation Advisory Board, whose expertise was so valuable in that aspect of our work.

- David Clarke, University of Melbourne

- Robert Davis, Rutgers University

- George Hein, Lesley College

- Mark St. John, Inverness Research Associates

# IMP National Advisory Board

Finally, we want to thank Steve Rasmussen, President of Key Curriculum Press, Casey FitzSimons, Key's Project Editor for the IMP curriculum, and the many others at Key whose work turned our ideas and words into published form.

*Dan Fendel     Dan Fendel     Lynne Alper     Sherry Fraser*

# *Foreword*

*Is There Really a Difference?* asks the title of one Year 2 unit of the Interactive Mathematics Program (IMP).

"You bet there is!" As Superintendent of Schools, I have found that IMP students in our District have more fun, are well prepared for our State Testing Program in the tenth grade, and are a more representative mix of the different groups in our geographical area than students in other pre-college math classes. Over the last few years, IMP has become an important example of curriculum reform in both our math and science programs.

When we decided in 1992 to pilot the Interactive Mathematics Program, we were excited about its modern approach to restructuring the traditional high school math sequence of courses and topics and its applied use of significant technology. We hoped that IMP would not only revitalize the pre-college math program, but also extend it to virtually all ninth-grade students. At the same time, we had a few concerns about whether IMP students would acquire all of the traditional course skills in algebra, geometry, and trigonometry.

Within the first year, the program proved successful and we were exceptionally pleased with the students' positive reaction and performance, the feedback from parents, and the enthusiasm of teachers. Our first group of IMP students, who graduated in June, 1996, scored as well on PSATs, SATs, and State tests as a comparable group of students in the traditional program did, and subsequent IMP groups are doing the same. In addition, the students have become our most enthusiastic and effective IMP promoters when visiting middle school classes to describe math course options to incoming ninth graders. One student commented, "IMP is the most fun math class I've ever had." Another said, "IMP makes you work hard, but you don't even notice it."

In our first pilot year, we found that the IMP course reached a broader range of students than the traditional Algebra 1 course did. It worked wonderfully not only for honors students, but for other students who would not have begun algebra study until tenth grade or later. The most successful students were those who became intrigued with exciting applications, enjoyed working in a group, and were willing to tackle the hard work of thinking seriously about math on a daily basis.

IMP Year 2 places the graphing calculator and computer in central positions early in the math curriculum. Students thrive on the regular group collaboration and grow in self-confidence and skill as they present their ideas to a large group. Most importantly, not only do students learn the symbolic and graphing applications of elementary algebra, the statistics of *Is There Really a Difference?,* and the geometry of *Do Bees Build It Best?,* but the concepts have meaning to them.

I wish you well as you continue your IMP path for a second year. I am confident that students and teachers using Year 2 will enjoy mathematics more than ever as they experiment, investigate, and discover solutions to the problems and activities presented this year.

Reginald Mayo
Superintendent
New Haven Public Schools
New Haven, Connecticut

# The Interactive Mathematics Program

### What is the Interactive Mathematics Program?

The Interactive Mathematics Program (IMP) is a growing collaboration of mathematicians, teacher-educators, and teachers who have been working together since 1989 on both curriculum development and professional development for teachers.

### What is the IMP curriculum?

IMP has created a four-year program of problem-based mathematics that replaces the traditional Algebra I–Geometry–Algebra II/Trigonometry–Precalculus sequence and that is designed to exemplify the curriculum reform called for in the *Curriculum and Evaluation Standards* of the National Council of Teachers of Mathematics (NCTM).

The IMP curriculum integrates traditional material with additional topics recommended by the NCTM *Standards,* such as statistics, probability, curve fitting, and matrix algebra. Although every IMP unit has a specific mathematical focus, most units are structured around a central problem and bring in other topics as needed to solve that problem, rather than narrowly restricting the mathematical content. Ideas that are developed in one unit are generally revisited and deepened in one or more later units.

### For which students is the IMP curriculum intended?

The IMP curriculum is for all students. One of IMP's goals is to make the learning of a core mathematics curriculum accessible to everyone. Toward that end, we have designed the program for use with heterogeneous classes. We provide you with a varied collection of supplemental problems to give you the flexibility to meet individual student needs.

*Teacher Phyllis Quick confers with a group of students.*

## How is the IMP classroom different?

When you use the IMP curriculum, your role changes from "imparter of knowledge" to observer and facilitator. You ask challenging questions. You do not give all the answers; rather, you prod students to do their own thinking, to make generalizations, and to go beyond the immediate problem by asking themselves "What if?" The IMP curriculum gives students many opportunities to write about their mathematical thinking, to reflect on what they have done, and to make oral presentations to one another about their work. In IMP, your assessment of students becomes integrated with learning, and you evaluate students according to a variety of criteria, including class participation, daily homework assignments, Problems of the Week, portfolios, and unit assessments. The IMP *Teaching Handbook* provides many practical suggestions on how to get the best possible results using this curriculum in *your* classroom.

## What is in Year 2 of the IMP curriculum?

Year 2 of the IMP curriculum contains five units.

### Solve It!

The focus of this unit is on using equations to represent real-life situations and on developing the skills to solve these equations. Students begin with situations used in the first year of the curriculum and develop algebraic representations of problems. In order to find solutions to the equations that occur, students explore the concepts of equivalent expressions and equivalent equations. Using these concepts, they develop principles such as the distributive property for working with algebraic expressions and equations, and they learn methods that they can use to solve any linear equation. They also explore the relationship between an algebraic expression, a function, an equation, and a graph, and they examine how to use graphs to solve nonlinear equations.

### Is There Really a Difference?

In this unit, students collect data and compare different population groups to one another. In particular, they concentrate on this question:

*If a sample from one population differs in some respect from a sample from a different population, how reliably can you infer that the overall populations differ in that respect?*

They begin by making double bar graphs of some classroom data and explore the process of making and testing hypotheses. Students realize that there is variation even among different samples from the same population, and they see the usefulness of the concept of a *null hypothesis* as they examine this variation. They build on their understanding of standard deviation from the Year 1 unit *The Pit and the Pendulum* and learn that the chi-square ($\chi^2$) statistic can give them the probability of seeing differences of a certain size in samples when the populations are really the same. Their work in this unit culminates in a two-week project in which they propose a hypothesis about two populations that they think really differ in some respect. They then collect sample data about the two populations and analyze their data by using bar graphs, tables, and the $\chi^2$ statistic.

*Do Bees Build It Best?*

In this unit students work on this problem:

> *Bees store their honey in honeycombs that consist of cells*
> *they make out of wax. What is the best design for a*
> *honeycomb?*

To analyze this problem, students begin by learning about area and the Pythagorean theorem. Then, using the Pythagorean theorem and trigonometry, students find a formula for the area of a regular polygon with fixed perimeter and find that the larger the number of sides, the larger the area of the polygon. Students then turn their attention to volume and surface area, focusing on prisms that have a regular polygon as the base. They find that for such prisms—if they also want the honeycomb cells to fit together—the mathematical winner, in terms of maximizing volume for a given surface area, is a regular hexagonal prism, which is essentially the choice of the bees.

*Cookies*

The focus of this unit is on graphing systems of linear inequalities and solving systems of linear equations. Although the central problem is one in linear programming, the major goal of the unit is for students to learn how to manipulate equations and how to reason using graphs.

Students begin by considering a classic type of linear programming problem in which they are asked to maximize the profits of a bakery that makes plain and iced cookies. They are constrained by the amount of ingredients they have on hand and the amount of oven and labor time available. First students work toward a graphical solution of the problem. They see how the linear function can be maximized or minimized by studying the graph. Because the maximum or minimum point they are looking for is often the intersection of two lines, they are motivated to investigate a method for solving two equations in two unknowns. They then return to work in groups on the cookie problem, each group presenting both a solution and a proof that their solution does maximize profits. Finally, each group invents its own linear programming problem and makes a presentation of the problem and its solution to the class.

### *All About Alice*

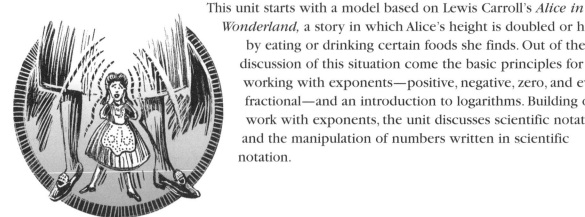

This unit starts with a model based on Lewis Carroll's *Alice in Wonderland,* a story in which Alice's height is doubled or halved by eating or drinking certain foods she finds. Out of the discussion of this situation come the basic principles for working with exponents—positive, negative, zero, and even fractional—and an introduction to logarithms. Building on the work with exponents, the unit discusses scientific notation and the manipulation of numbers written in scientific notation.

## *How do the four years of the IMP curriculum fit together?*

The four years of the IMP curriculum form an integrated sequence through which students can learn the mathematics they will need both for further education and on the job. Although the organization of the IMP curriculum is very different from the traditional Algebra I–Geometry–Algebra II/Trigonometry–Precalculus sequence, the important mathematical ideas are all there.

Here are some examples of how both traditional concepts and topics new to the high school curriculum are developed:

### Linear equations

In Year 1 of the IMP curriculum, students develop an intuitive foundation of algebraic thinking, including the use of variables, which they build on throughout the program. In the Year 2 unit *Solve It!,* students use the concept of equivalent equations to see how to solve any linear equation in a single variable. Later in Year 2, in a unit called *Cookies* (about maximizing profits for a bakery), they solve pairs of linear equations in two variables, using both algebraic and geometric methods. In *Meadows or Malls?* (Year 3), they extend those ideas to systems with more than two variables, and see how to use matrices and the technology of graphing calculators to solve such systems.

### Measurement and the Pythagorean theorem

Measurement, including area and volume, is one of the fundamental topics in geometry. The Pythagorean theorem is one of the most important geometric principles ever discovered. In the Year 2 unit *Do Bees Build It Best?,* students combine these ideas with their knowledge of similarity (from the Year 1 unit *Shadows*) to see why the hexagonal prism of the bees' honeycomb design is the most efficient regular prism possible. Students also use the Pythagorean theorem in later units, applying it to develop such principles as the distance formula in coordinate geometry.

### Trigonometric functions

In traditional programs, the trigonometric functions are introduced in the eleventh or twelfth grade. In the IMP curriculum, students begin working with trigonometry in Year 1 (in *Shadows*), using right-triangle trigonometry in several units in Years 2 and 3 (including *Do Bees Build It Best?*). In the Year 4 unit *High Dive,* they extend trigonometry from right triangles to circular functions, in the context of a circus act in which a performer falls from a Ferris wheel into a moving tub of water. (In *High Dive,* students also learn principles of physics, developing laws for falling objects and using vectors to find vertical and horizontal components of velocity.)

### Standard deviation and the binomial distribution

Standard deviation and the binomial distribution are major tools in the study of probability and statistics. *The Game of Pig* gets students started by building a firm understanding of concepts of probability and the phenomenon of experimental variation. Later in Year 1 (in *The Pit and the Pendulum*), they use standard deviation to see that the period of a pendulum is determined primarily by its length. In Year 2, they compare standard deviation with the chi-square test in examining whether the difference between two sets of data is statistically significant. In *Pennant Fever* (Year 3), students use the binomial distribution to evaluate a team's chances of winning the baseball championship, and in *The Pollster's Dilemma* (Year 4), students tie many of these ideas together in the central limit theorem, seeing how the margin of error and the level of certainty for an election poll depend on its size.

## *Does the program work?*

The IMP curriculum has been thoroughly field-tested and enthusiastically received by hundreds of classroom teachers around the country. Their enthusiasm is based on the success they have seen in their own classrooms with their own students. For instance, IMP teacher Dennis Cavaillé says, "For the first time in my teaching career, I see lots of students excited about solving math problems inside *and* outside of class."

These informal observations are backed up by more formal evaluations. Dr. Norman Webb of the Wisconsin Center for Education Research has done several studies comparing the performance of students using the IMP curriculum with the performance of students in traditional programs. For instance, he has found that IMP students do as well as students in traditional mathematics classes on standardized tests such as the SAT. This is especially significant because IMP students spend about twenty-five percent of their time studying topics, such as statistics, not covered on these tests. To measure IMP students' achievement in these other areas, Dr. Webb conducted three separate studies involving students at different grade levels and in different locations. The three tests used in these studies involved statistics, quantitative reasoning, and general problem solving. In all three cases, the IMP students outperformed their counterparts in traditional programs by a statistically significant margin, even though the two groups began with equivalent scores on eighth grade standardized tests.

But one of our proudest achievements is that IMP students are excited about mathematics, as shown by Dr. Webb's finding that they take more mathematics courses in high school than their counterparts in traditional programs. We think this is because they see that mathematics can be relevant to their own lives. If so, then the program works.

Dan Fendel

Diane Resek

Lynne Alper

Sherry Fraser

# Note to Students

*These pages in the student book welcome students to the program.*

This textbook represents the second year of a four-year program of mathematics learning and investigation. As in the first year, the program is organized around interesting, complex problems, and the concepts you learn grow out of what you'll need to solve those problems.

## • *If you studied IMP Year 1*

If you studied IMP Year 1, then you know the excitement of problem-based mathematical study, such as devising strategies for a complex dice game, learning the history of the Overland Trail, and experimenting with pendulums.

The Year 2 program extends and expands the challenges that you worked with in Year 1. For instance:

- In Year 1, you began developing a foundation for working with variables. In Year 2, you will build on this foundation in units that demonstrate the power of algebra to solve problems, including some that look back at situations from Year 1 units.

- In Year 1, you used principles of statistics to help predict the period of a 30-foot pendulum. In Year 2, you will learn another statistical method, one that will help you to understand statistical comparisons of populations. One important part of your work will be to prepare, conduct, and analyze your own survey.

You'll also use ideas from geometry to understand why the design of bees' honeycombs is so efficient, and you'll use graphs to help a bakery decide how many plain and iced cookies they should make to maximize their profits. Year 2 closes with a literary adventure—you'll use Lewis Carroll's classic *Alice's Adventures in Wonderland* to explore and extend the meaning of exponents.

• *If you didn't study IMP Year 1*

If this is your first experience with the Interactive Mathematics Program (IMP), you can rely on your classmates and your teacher to fill in what you've missed. Meanwhile, here are some things you should know about the program, how it was developed, and how it is organized.

The Interactive Mathematics Program is the product of a collaboration of teachers, teacher-educators, and mathematicians who have been working together since 1989 to reform the way high school mathematics is taught. About one hundred thousand students and five hundred teachers used these materials before they were published. Their experiences, reactions, and ideas have been incorporated into this final version.

Our goal is to give you the mathematics you need in order to succeed in this changing world. We want to present mathematics to you in a manner that reflects how mathematics is used and that reflects the different ways people work and learn together. Through this perspective on mathematics, you will be prepared both for continued study of mathematics in college and for the world of work.

This book contains the various assignments that will be your work during Year 2 of the program. As you will see, these problems require ideas from many branches of mathematics, including algebra, geometry, probability, graphing, statistics, and trigonometry. Rather than present each of these areas separately, we have integrated them and presented them in meaningful contexts, so you will see how they relate to each other and to our world.

Each unit in this four-year program has a central problem or theme, and focuses on several major mathematical ideas. Within each unit, the material is organized for teaching purposes into "days," with a homework assignment for each day. (Your class may not follow this schedule exactly, especially if it doesn't meet every day.)

At the end of the main material for each unit, you will find a set of supplementary problems. These problems provide you with additional opportunities to work with ideas from the unit, either to strengthen your understanding of the core material or to explore new ideas related to the unit.

Although the IMP program is not organized into courses called "Algebra," "Geometry," and so on, you will be learning all the essential mathematical concepts that are part of those traditional courses. You will also be learning concepts from branches of mathematics—especially statistics and probability—that are not part of a traditional high school program.

To accomplish your goals, you will have to be an active learner, because the book does not teach directly. Your role as a mathematics student will be to experiment, to investigate, to ask questions, to make and test conjectures, and to reflect, and then to communicate your ideas and conclusions both orally and in writing. You will do some of your work in collaboration with fellow students, just as users of mathematics in the real world often work in teams. At other times, you will be working on your own.

We hope you will enjoy the challenge of this new way of learning mathematics and will see mathematics in a new light.

*Dan Fendel   Diane Resek   Lynne Alper   Sherry Fraser*

# *Finding What You Need*

We designed this guide to help you find what you need amid all the information it provides. Each of the following components has a special treatment in the layout of the guide.

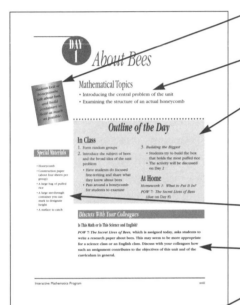

**Synopsis of the Day:** The key idea or activity for each day is summarized in a brief sentence or two.

**Mathematical Topics:** Mathematical issues for the day are presented in a bulleted list.

**Outline of the Day:** Under the *In Class* heading, the outline summarizes the activities for the day, which are keyed to numbered headings in the discussion. Daily homework assignments and Problems of the Week are listed under the *At Home* heading.

**Special Materials Needed:** Special items needed in the classroom for each day are bulleted here.

**Discuss With Your Colleagues:** This section highlights topics that you may want to discuss with your peers.

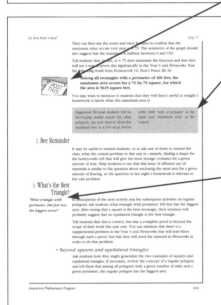

**Post This:** The *Post This* icon indicates items that you may want to display in the classroom.

**Asides:** These are ideas outside the main thrust of a discussion. They include background information, refinements, or subtle points that may only be of interest to some students, ways to help fill in gaps in understanding the main ideas, and suggestions about when to bring in a particular concept.

**Suggested Questions:** These are specific questions that you might ask during an activity or discussion to promote student insight or to determine whether students understand an idea. The appropriateness of these questions generally depends on what students have already developed or presented on their own.

## Icons for Student Written Products

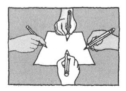

***Single Group report***

***Individual reports***

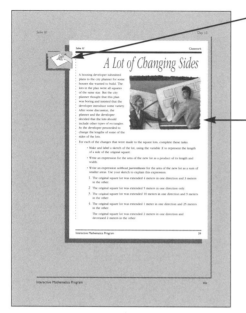

**Icons for Student Written Products:** For each group activity, there is an icon suggesting a single group report, individual reports, or no report at all. If graphs are included, the icon indicates this as well. (The graph icons do not appear in every unit.)

**Embedded Student Pages:** The teacher guide contains reduced-size copies of the pages from the student book, including the "transition pages" that appear occasionally within each unit to summarize each portion of the unit and to prepare students for what is coming. The reduced-size classwork and homework assignments follow the teacher notes for the day on which the activity is begun. Having all of these student pages in the teacher's guide is a helpful way for you to see things from the students' perspective.

# Additional Information

Here is a brief outline of other tools we have included to assist you and make both the teaching and the learning experience more rewarding.

**Glossary:** This section, which is found at the back of the book, gives the definitions of important terms for all of Year 2 for easy reference. The same glossary appears in the student book.

**Appendix A: Supplemental Problems:** This appendix contains a variety of interesting additional activities for the unit, for teachers who would like to supplement material found in the regular classroom problems. These additional activities are of two types—*reinforcements,* which help increase student understanding of concepts that are central to the unit, and *extensions,* which allow students to explore ideas beyond the basic unit.

**Appendix B: Blackline Masters:** For each unit, this appendix contains materials you can reproduce that are not available in the student book and that will be helpful to teacher and student alike. They include the end-of-unit assessments as well as such items as diagrams from which you can make transparencies. Semester assessments for Year 2 are included in *Do Bees Build It Best?* (for first semester) and *All About Alice* (for second semester).

**Single group graph**        **Individual graphs**        **No report at all**

# Year 2 IMP Units

*Solve It!*       (in this book)

*Is There Really a Difference?*

*Do Bees Build It Best?*

*Cookies*

*All About Alice*

# Contents

*Solve It!* Overview . . . . . . . . . . . . . . . . . . . . . . . . . . . . . . . . . . . .xxvii

   Summary of the Unit . . . . . . . . . . . . . . . . . . . . . . . . . . . . . . . .xxvii

   Concepts and Skills . . . . . . . . . . . . . . . . . . . . . . . . . . . . . . . .xxviii

   Materials . . . . . . . . . . . . . . . . . . . . . . . . . . . . . . . . . . . . . . . . .xxix

   Grading . . . . . . . . . . . . . . . . . . . . . . . . . . . . . . . . . . . . . . . . . .xxx

   Days 1–5: Solving Equations and Understanding Situations
   (Reduced Student Page) . . . . . . . . . . . . . . . . . . . . . . . . . . . . . .2

Day 1: Setting the Tone . . . . . . . . . . . . . . . . . . . . . . . . . . . . . . . . . .3

   Forming Groups . . . . . . . . . . . . . . . . . . . . . . . . . . . . . . . . . . . . .3

   Overview of the Year: *A Year 2 Sampler* . . . . . . . . . . . . . . . . . . . .3

   Getting to Know You . . . . . . . . . . . . . . . . . . . . . . . . . . . . . . . . .4

   Reflections on Year 1 . . . . . . . . . . . . . . . . . . . . . . . . . . . . . . . . .5

   *Homework 1: Math, Me, and the Future* . . . . . . . . . . . . . . . . . . .7

Day 2: Memories of Yesteryear . . . . . . . . . . . . . . . . . . . . . . . . . . . .11

   Discussion of *Homework 1: Math, Me, and the Future* . . . . . . . . . .12

   Introduction to *Solve It!* . . . . . . . . . . . . . . . . . . . . . . . . . . . . .12

   *Memories of Yesteryear* . . . . . . . . . . . . . . . . . . . . . . . . . . . . . .13

   *Homework 2: Building a Foundation* . . . . . . . . . . . . . . . . . . . . .14

Day 3: Memories of Yesteryear, Continued . . . . . . . . . . . . . . . . . . .19

   Discussion of *Homework 2: Building a Foundation* . . . . . . . . . . . .20

   Discussion of *Memories of Yesteryear* . . . . . . . . . . . . . . . . . . . . .22

   Introduction to *Homework 3: You're the Storyteller* . . . . . . . . . . . .23

   *Homework 3: You're the Storyteller* . . . . . . . . . . . . . . . . . . . . . .23

Day 4: Is It a Digit? . . . . . . . . . . . . . . . . . . . . . . . . . . . . . . . . . . . .25

   Discussion of *Homework 3: You're the Storyteller* . . . . . . . . . . . . .25

   *Is It a Digit?* . . . . . . . . . . . . . . . . . . . . . . . . . . . . . . . . . . . . . .26

   Discussion of *Is It a Digit?* . . . . . . . . . . . . . . . . . . . . . . . . . . . .26

   Introduction to *POW 1: A Digital Proof* . . . . . . . . . . . . . . . . . . .27

   For Reference: *The Standard POW Write-up* . . . . . . . . . . . . . . . . .27

   *Homework 4: Running on the Overland Trail* . . . . . . . . . . . . . . . .28

Day 5: Lamppost Shadows . . . . . . . . . . . . . . . . . . . . . . . . . . . . . . . . .35

   Discussion of *Homework 4: Running on the Overland Trail* . . . . . . . . . . .35

   *Lamppost Shadows* . . . . . . . . . . . . . . . . . . . . . . . . . . . . . . . . . . . .37

   *Homework 5: 1-2-3-4 Puzzle with Negatives* . . . . . . . . . . . . . . . . . . . . .37

      Days 6–12: Keeping Things Balanced (Reduced Student Page) . . . . . . . . .41

Day 6: Similarity Equations . . . . . . . . . . . . . . . . . . . . . . . . . . . . . . . .43

   Discussion of *Homework 5: 1-2-3-4 Puzzle with Negatives* . . . . . . . . . . . .43

   Discussion of *Lamppost Shadows* . . . . . . . . . . . . . . . . . . . . . . . . . . .44

   Introduction to *Homework 6: The Mystery Bags Game* . . . . . . . . . . . . . .46

   *Homework 6: The Mystery Bags Game* . . . . . . . . . . . . . . . . . . . . . . . .47

Day 7: Equations for Bags . . . . . . . . . . . . . . . . . . . . . . . . . . . . . . . . .51

   Discussion of *Homework 6: The Mystery Bags Game* . . . . . . . . . . . . . . .51

   Representing the Mystery Bags Algebraically . . . . . . . . . . . . . . . . . . . .53

   *Homework 7: You're the Jester* . . . . . . . . . . . . . . . . . . . . . . . . . . . .53

Day 8: Substitution and Order of Operations Revisited . . . . . . . . . . . . .55

   Discussion of *Homework 7: You're the Jester* . . . . . . . . . . . . . . . . . . .55

   Reviewing Substitution . . . . . . . . . . . . . . . . . . . . . . . . . . . . . . . . .56

   For Reference: *Substitution and Evaluation* . . . . . . . . . . . . . . . . . . . .57

   *Homework 8: Letters, Numbers, and a Story* . . . . . . . . . . . . . . . . . . . .58

Day 9: Catching Up . . . . . . . . . . . . . . . . . . . . . . . . . . . . . . . . . . . . .63

   Discussion of *Homework 8: Letters, Numbers, and a Story* . . . . . . . . . . .64

   *Catching Up* . . . . . . . . . . . . . . . . . . . . . . . . . . . . . . . . . . . . . . .64

   Discussion of *Catching Up* . . . . . . . . . . . . . . . . . . . . . . . . . . . . . .64

   *Homework 9: More Letters, Numbers, and Mystery Bags* . . . . . . . . . . . .66

Day 10: Back to the Lake . . . . . . . . . . . . . . . . . . . . . . . . . . . . . . . . .71

   POW Presentation Preparation . . . . . . . . . . . . . . . . . . . . . . . . . . . .72

   Discussion of *Homework 9: More Letters, Numbers, and Mystery Bags* . . .72

   *Back to the Lake* . . . . . . . . . . . . . . . . . . . . . . . . . . . . . . . . . . . . .72

   Discussion of *Back to the Lake* . . . . . . . . . . . . . . . . . . . . . . . . . . . .73

   Review of Function Notation . . . . . . . . . . . . . . . . . . . . . . . . . . . . . .74

   *Homework 10: What Will It Answer?* . . . . . . . . . . . . . . . . . . . . . . . . .75

## Day 11: POW 1 Presentations . . . . . . . . . . . . . . . . . . . . . . . . . .81

Forming New Groups . . . . . . . . . . . . . . . . . . . . . . . . . . .81

Discussion of *Homework 10: What Will It Answer?* . . . . . . . . . . . . . . . . .82

Presentations of *POW 1: A Digital Proof* . . . . . . . . . . . . . . . . . . . . . .83

Introduction to *POW 2: Tying the Knots* . . . . . . . . . . . . . . . . . . . .84

*Homework 11: Line It Up* . . . . . . . . . . . . . . . . . . . . . . . . . . .84

## Day 12: Using a Graph . . . . . . . . . . . . . . . . . . . . . . . . . . . . .89

Discussion of *Homework 11: Line It Up* . . . . . . . . . . . . . . . . . . . . .90

Review of Graphing on the Graphing Calculator . . . . . . . . . . . . . . . . .91

*The Graph Solves the Problem* . . . . . . . . . . . . . . . . . . . . . . . . .92

*Homework 12: Who's Right?* . . . . . . . . . . . . . . . . . . . . . . . . . . .93

Days 13–21: What's the Same? (Reduced Student Page) . . . . . . . . . . . . . . .96

## Day 13: A Lot About Lots . . . . . . . . . . . . . . . . . . . . . . . . . . .97

Discussion of *Homework 12: Who's Right?* . . . . . . . . . . . . . . . . . . . .98

*A Lot of Changing Sides* . . . . . . . . . . . . . . . . . . . . . . . . . . .100

*Homework 13: Why Are They Equivalent?* . . . . . . . . . . . . . . . . . . . .102

## Day 14: Lots More . . . . . . . . . . . . . . . . . . . . . . . . . . . . . .107

Discussion of *Homework 13: Why Are They Equivalent?* . . . . . . . . . . . . .107

Discussion of *A Lot of Changing Sides* . . . . . . . . . . . . . . . . . . . . .108

*Homework 14: One Each Way* . . . . . . . . . . . . . . . . . . . . . . . . .110

## Day 15: Distributing the Area . . . . . . . . . . . . . . . . . . . . . . . .113

POW Presentation Preparation . . . . . . . . . . . . . . . . . . . . . . . .114

Discussion of *Homework 14: One Each Way* . . . . . . . . . . . . . . . . . .114

*Distributing the Area* . . . . . . . . . . . . . . . . . . . . . . . . . . . .114

Discussion of *Distributing the Area* . . . . . . . . . . . . . . . . . . . . . .114

The Distributive Property: Officially! . . . . . . . . . . . . . . . . . . . . .115

*Homework 15: The Distributive Property and Mystery Lots* . . . . . . . . . .116

## Day 16: POW 2 Presentations and Solidifying the Distributive Property . . . . .121

Presentations of *POW 2: Tying the Knots* . . . . . . . . . . . . . . . . . . .122

Discussion of *Homework 15: The Distributive Property and Mystery Lots* 123

Ordinary Multiplication and the Distributive Property . . . . . . . . . . . . .124

*Homework 16: Views of the Distributive Property* . . . . . . . . . . . . . . .124

## Day 17: The Distributive Property and Divisor Counting . . . . . . . . . . . . . . . .129

Discussion of *Homework 16: Views of the Distributive Property* . . . . . . .130

*Prime Time* . . . . . . . . . . . . . . . . . . . . . . . . . . . . . . . . . . . . . . . . . . . . .131

Discussion of *Prime Time* . . . . . . . . . . . . . . . . . . . . . . . . . . . . . . . . . . .132

Introduction of *POW 3: Divisor Counting* . . . . . . . . . . . . . . . . . . . . . . .132

*Homework 17: Exactly Three or Four* . . . . . . . . . . . . . . . . . . . . . . . . . .133

## Day 18: Taking Some Out . . . . . . . . . . . . . . . . . . . . . . . . . . . . . . . . . . . . .139

Discussion of *Homework 17: Exactly Three or Four* . . . . . . . . . . . . . . .140

*Taking Some Out, Part I* . . . . . . . . . . . . . . . . . . . . . . . . . . . . . . . . . . . .140

Discussion of *Taking Some Out, Part I* . . . . . . . . . . . . . . . . . . . . . . . . .140

*Homework 18: Subtracting Some Sums* . . . . . . . . . . . . . . . . . . . . . . . .141

## Day 19: Taking Some More Out . . . . . . . . . . . . . . . . . . . . . . . . . . . . . . . .147

Discussion of *Homework 18: Subtracting Some Sums* . . . . . . . . . . . . . .147

*Taking Some Out, Part II* . . . . . . . . . . . . . . . . . . . . . . . . . . . . . . . . . . .149

*Homework 19: Randy, Sandy, and Dandy Return* . . . . . . . . . . . . . . . .149

## Day 20: Equivalent Equations . . . . . . . . . . . . . . . . . . . . . . . . . . . . . . . . .153

Discussion of Part I of *Homework 19: Randy, Sandy, and Dandy Return* . .153

Discussion of *Taking Some Out, Part II* . . . . . . . . . . . . . . . . . . . . . . . .154

Discussion of Part II of *Homework 19: Randy, Sandy, and Dandy Return* 156

Equivalent Equations . . . . . . . . . . . . . . . . . . . . . . . . . . . . . . . . . . . . . . .156

*Homework 20: Equation Time* . . . . . . . . . . . . . . . . . . . . . . . . . . . . . . .158

## Day 21: Scrambling Equations . . . . . . . . . . . . . . . . . . . . . . . . . . . . . . . . .161

Forming New Groups . . . . . . . . . . . . . . . . . . . . . . . . . . . . . . . . . . . . . . .161

Discussion of *Homework 20: Equation Time* . . . . . . . . . . . . . . . . . . . .162

*Scrambling Equations* . . . . . . . . . . . . . . . . . . . . . . . . . . . . . . . . . . . . . .162

*Homework 21: More Scrambled Equations and Mystery Bags* . . . . . . . . .164

Days 22–26: The Linear World (Reduced Student Page) . . . . . . . . . . . . . .169

## Day 22: Summarizing Linear Equations . . . . . . . . . . . . . . . . . . . . . . . . . .171

POW Presentation Preparation . . . . . . . . . . . . . . . . . . . . . . . . . . . . . . . .171

Discussion of *Homework 21: More Scrambled Equations
and Mystery Bags* . . . . . . . . . . . . . . . . . . . . . . . . . . . . . . . . . . . . . . . . .172

Linear Equations . . . . . . . . . . . . . . . . . . . . . . . . . . . . . . . . . . . . . . . . . .172

*Old Friends and New Friends* . . . . . . . . . . . . . . . . . . . . . . . . . . . . . . . .173

*Homework 22: New Friends Visit Your Home* . . . . . . . . . . . . . . . . . . . . .174

## Day 23: POW 3 Presentations . . . . . . . . . . . . . . . . . . . . . . .179

Discussion of *Homework 22: New Friends Visit Your Home* . . . . . . . . . . .179

Discussion of *Old Friends and New Friends* . . . . . . . . . . . . . . . . . . . . . .180

Presentations of *POW 3: Divisor Counting* . . . . . . . . . . . . . . . . . . . . . .180

*Homework 23: From One Variable to Two* . . . . . . . . . . . . . . . . . . . . . . .181

## Day 24: Get It Straight . . . . . . . . . . . . . . . . . . . . . . . . . . . . .183

Discussion of *Homework 23: From One Variable to Two* . . . . . . . . . . . .184

*Get It Straight* . . . . . . . . . . . . . . . . . . . . . . . . . . . . . . . . . . . . . . . . . .185

*Homework 24: A Distributive Summary* . . . . . . . . . . . . . . . . . . . . . . . .185

## Day 25: Linear Functions and Straight-Line Graphs . . . . . . .189

Discussion of *Homework 24: A Distributive Summary* . . . . . . . . . . . . . .189

Continued Work on *Get It Straight* . . . . . . . . . . . . . . . . . . . . . . . . . . . .189

*Homework 25: All by Itself* . . . . . . . . . . . . . . . . . . . . . . . . . . . . . . . . . .190

## Day 26: Get It Straight Presentations and Variable Solutions . . . . . . .193

Presentations of *Get It Straight* . . . . . . . . . . . . . . . . . . . . . . . . . . . . . .193

Discussion of *Homework 25: All by Itself* . . . . . . . . . . . . . . . . . . . . . . .194

*Homework 26: More Variable Solutions* . . . . . . . . . . . . . . . . . . . . . . . .195

Days 27–31: Beyond Linearity (Reduced Student Page) . . . . . . . . . . . . .198

## Day 27: Where's Speedy? . . . . . . . . . . . . . . . . . . . . . . . . . . .199

Discussion of *Homework 26: More Variable Solutions* . . . . . . . . . . . . . .199

Beyond Linear Equations . . . . . . . . . . . . . . . . . . . . . . . . . . . . . . . . . . . .200

*Where's Speedy?* . . . . . . . . . . . . . . . . . . . . . . . . . . . . . . . . . . . . . . . . .201

Discussion of *Where's Speedy?* . . . . . . . . . . . . . . . . . . . . . . . . . . . . . . .201

*Homework 27: A Mixed Bag* . . . . . . . . . . . . . . . . . . . . . . . . . . . . . . . . .201

## Day 28: Graphs and Equations . . . . . . . . . . . . . . . . . . . . . .205

Discussion of *Homework 27: A Mixed Bag* . . . . . . . . . . . . . . . . . . . . . .205

*To the Rescue* . . . . . . . . . . . . . . . . . . . . . . . . . . . . . . . . . . . . . . . . . . .206

Discussion of *To the Rescue* . . . . . . . . . . . . . . . . . . . . . . . . . . . . . . . . .206

*Homework 28: Swinging Pendulum* . . . . . . . . . . . . . . . . . . . . . . . . . . .207

Day 29: Mystery Graph . . . . . . . . . . . . . . . . . . . . . . . . . . . . . . . . . . . . . . . . .211

    Discussion of *Homework 28: Swinging Pendulum* . . . . . . . . . . . . . . .211

    *Mystery Graph* . . . . . . . . . . . . . . . . . . . . . . . . . . . . . . . . . . . . . . . . .212

    Discussion of *Mystery Graph* . . . . . . . . . . . . . . . . . . . . . . . . . . . . . . .213

    *Homework 29: Functioning in the Math World* . . . . . . . . . . . . . . . .214

Day 30: A Graphing Calculator Approach . . . . . . . . . . . . . . . . . . . . . . . .219

    Discussion of *Homework 29: Functioning in the Math World* . . . . . . .219

    *The Graphing Calculator Solver* . . . . . . . . . . . . . . . . . . . . . . . . . . . .220

    *Homework 30: A Solving Sampler* . . . . . . . . . . . . . . . . . . . . . . . . . . .221

Day 31: Portfolios . . . . . . . . . . . . . . . . . . . . . . . . . . . . . . . . . . . . . . . . . . . .225

    Discussion of *Homework 30: A Solving Sampler* . . . . . . . . . . . . . . . .225

    *"Solve It!" Portfolio* . . . . . . . . . . . . . . . . . . . . . . . . . . . . . . . . . . . . .225

    Homework: Prepare for Assessments . . . . . . . . . . . . . . . . . . . . . . . . .226

Day 32: Final Assessments . . . . . . . . . . . . . . . . . . . . . . . . . . . . . . . . . . . .229

    End-of-Unit Assessments . . . . . . . . . . . . . . . . . . . . . . . . . . . . . . . . . .229

Day 33: Summing Up . . . . . . . . . . . . . . . . . . . . . . . . . . . . . . . . . . . . . . . .233

    Discussion of Unit Assessments . . . . . . . . . . . . . . . . . . . . . . . . . . . . .233

    Unit Summary . . . . . . . . . . . . . . . . . . . . . . . . . . . . . . . . . . . . . . . . . .234

Appendix A: Supplemental Problems . . . . . . . . . . . . . . . . . . . . . . . . . . . .235

    Appendix: Supplemental Problems (Reduced Student Page) . . . . . . . . .237

    *What to Expect?* . . . . . . . . . . . . . . . . . . . . . . . . . . . . . . . . . . . . . . . .238

    *Carlos and Betty* . . . . . . . . . . . . . . . . . . . . . . . . . . . . . . . . . . . . . . .239

    *Ten Missing Digits* . . . . . . . . . . . . . . . . . . . . . . . . . . . . . . . . . . . . . .240

    *Same Expectations* . . . . . . . . . . . . . . . . . . . . . . . . . . . . . . . . . . . . . .241

    *Preserve the Distributive Property* . . . . . . . . . . . . . . . . . . . . . . . . . .242

    *The Locker Problem* . . . . . . . . . . . . . . . . . . . . . . . . . . . . . . . . . . . . .243

    *Who's Got an Equivalent?* . . . . . . . . . . . . . . . . . . . . . . . . . . . . . . . .244

    *Make It Simple* . . . . . . . . . . . . . . . . . . . . . . . . . . . . . . . . . . . . . . . . .245

    *Linear in a Variable* . . . . . . . . . . . . . . . . . . . . . . . . . . . . . . . . . . . .246

    *The Shadow Equation Revisited* . . . . . . . . . . . . . . . . . . . . . . . . . . . .247

    *A Function—Not!* . . . . . . . . . . . . . . . . . . . . . . . . . . . . . . . . . . . . . . .248

Appendix B: Blackline Masters . . . . . . . . . . . . . . . . . . . . . . . . . . . . . . . . .251

Glossary . . . . . . . . . . . . . . . . . . . . . . . . . . . . . . . . . . . . . . . . . . . . . . . . . . .257

# *"Solve It!" Overview*

## *Summary of the Unit*

The focus of this unit is on using equations to represent real-life situations and on developing the skills needed to solve these equations in different ways. Students begin with situations they encountered in the first year of the curriculum and develop algebraic representations of various problems. During this part of the unit, they review some Year 1 concepts, such as negative numbers, functions, probability, and similar triangles.

After this work on representing situations with equations, students are introduced to a game involving a pan balance. They see how to represent the game algebraically, and the pan-balance model becomes a useful metaphor throughout the unit for working with equations. At this stage of the unit, students' work with equations is kept at an intuitive level; it is formalized only after they have become more comfortable with equations.

After a brief review of substitution and the conventions for order of operations, students look at the use of graphs of functions for solving equations. They see that a graph can be used to solve an equation but that this approach often gives only approximate solutions.

The unit then turns to a more algebraic approach, beginning with the concept of **equivalent expressions.** As part of this work, students use an area model for multiplication as a tool for understanding the distributive property. Because of its importance in algebraic manipulations, the distributive property is discussed from other perspectives as well.

A physical model (the hot-and-cold-cube model from Year 1) is also used to help students understand the subtraction of expressions in parentheses and to develop more tools for simplifying algebraic expressions.

With some algebraic concepts and techniques in hand, students next apply these ideas to equation solving. They formalize their earlier work with the pan-balance model and generalize their ideas using the concept of **equivalent equations.** They see that they can use algebraic methods to get the exact solution to any linear equation in one variable. They then explore linear functions and their graphs and work on solving equations for one variable in terms of others.

The unit concludes with a look at how graphical methods can be used to solve nonlinear equations. Students see that a graph can be thought of as an answer key to a whole family of equations. They also work with graphs abstractly, draw graphs that fit various conditions, and answer questions about functions from their graphs when they have no algebraic expression for the function.

This list summarizes the unit's overall organization.

- Day 1: Introduction to Year 2 of IMP

- Days 2–6: Introduction to the use of equations to solve problems and review of negative numbers

- Days 7–8: Introduction of the pan-balance metaphor and review of substitution and order of operations

- Days 9–12: Review of functions, function notation, and graphs of functions

- Days 13–17: Equivalent expressions and the distributive property

- Days 18–20: Subtraction of expressions in parentheses

- Days 21–23: Equivalent equations and algebraic methods for solving linear equations

- Days 24–26: Investigation of the graphs of linear functions

- Days 27–30: Use of functions to develop equations and of graphs to solve equations

- Days 31–33: Portfolios, assessments, and summary

## Concepts and Skills

The focus of this unit is on the basic principles of algebraic manipulation—equivalent expressions and equivalent equations—and the development of algebraic and other techniques for solving equations. The main concepts and skills that students will encounter and practice during the course of the unit can be summarized by category.

### Algebraic representation of problem situations

- Expressing questions about familiar situations using equations

- Using a guess-and-check approach to problem solving as a tool for developing equations

- Reviewing substitution and conventions for order of operations

### General concepts about functions

- Understanding the relationship between an algebraic rule, a table, and a graph

- Reviewing function notation

- Recognizing that certain equations lead to straight-line graphs

- Representing real-life situations using functions

*Equivalent equations*

- Using a pan balance as a metaphor for working with equations

- Developing principles for creating equations that are equivalent to a given equation

*Equivalent expressions and the distributive property*

- Recognizing that different algebraic expressions can give the same numerical value for all substitutions

- Using the area of a rectangle as a model for multiplication

- Using the definition of multiplication as repeated addition as a way of understanding the distributive property

- Using numerical examples to gain insight into the distributive property

- Making a formal statement of the distributive property

- Seeing the role of the distributive property in the standard multiplication algorithm

- Using a physical model to understand the subtraction of expressions in parentheses

*Linear equations and functions*

- Identifying expressions of the form $ax + b$ as defining linear functions

- Developing techniques for algebraically solving linear equations in one variable

- Examining equations with several variables that are linear in a given variable and solving for that variable in terms of others

- Investigating how the coefficient of $x$ and the constant term of a linear function affect its graph

*Solving equations graphically*

- Defining functions using expressions from equations

- Solving equations using the graphs of related functions

- Using graphing calculators to solve a variety of equations

Other topics will arise in the discussion of the Problems of the Week.

## Materials

Here at the start of Year 2, we remind you of the standard materials you will need to provide.

- A class set of graphing calculators

- An overhead projector and screen, blank transparencies, and pens for transparencies

- A device for overhead projection of the graphing calculator screen

- A deck of cards for random grouping of students

- Wall space for posting student work

- A blank attendance chart (a classroom layout of desks that can be filled in each time groups change)

- Boxes or crates with folders for student portfolios

- Tape, glue, construction paper, paper clips, and pencils

- Pads of $2' \times 3'$ paper or rolls of butcher paper for posters

- A set of felt markers for each group

- Rulers, protractors, and scissors

- Bags or other containers to hold materials for each group

In addition, you will need these materials for this particular unit.

- (Optional) Student portfolios from Year 1

- String (for a simulation of POW 2)

- A transparency of a graph for Day 29 (see Appendix B)

For a list of materials that students need to provide, see the subsection "Materials" on Day 1.

## Grading

The IMP *Teaching Handbook* contains general ideas about how to grade students in an IMP class. You will probably want to check daily that students have done their homework and include regularity of homework completion as part of students' grades. Your grading scheme will probably also include Problems of the Week, the unit portfolio, and the end-of-unit assessments.

Because you will not be able to read thoroughly every assignment that students turn in, you will need to select certain assignments to read carefully and to base grades on. Here are some suggestions.

- *Memories of Yesteryear* (Days 2 and 3) or *Homework 4: Running on the Overland Trail*

- *Homework 9: More Letters, Numbers, and Mystery Bags*

- *Homework 14: One Each Way*

- *Homework 22: New Friends Visit Your Home*

- *Get It Straight* (Days 24–26)

- *Mystery Graph* (Day 29)

If you want to base your grading on more tasks, there are many other homework assignments, class activities, and oral presentations you can use.

# Interactive Mathematics Program

**IMP**

Integrated High School Mathematics

YEAR **2**

*Solve It!*

*Solve It!*

**Days 1-5**

# Solving Equations and Understanding Situations

For many people, mathematics means solving equations. You will, in fact, solve lots of equations in this first unit of Year 2. But solving equations makes more sense if those equations describe something meaningful.

*This page in the student book introduces Days 1 through 5.*

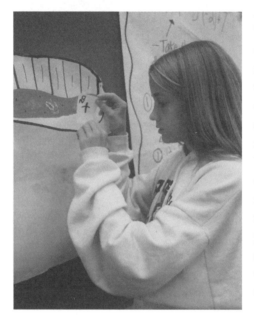

*Tiffany Bock dramatically demonstrates to the class a model for understanding the arithmetic of positive and negative numbers.*

After a look back at Year 1, this unit takes up the challenge of representing situations using algebra and equations. Some of the situations you will study may be familiar from Year 1. The opening days of this unit also include some review of the arithmetic of positive and negative numbers, in the context of a Year 1 Problem of the Week.

# DAY 1
# Setting the Tone

## Mathematical Topics

• Looking back at Year 1 of IMP

## Outline of the Day

### In Class

1. Form new random groups
2. Introduce Year 2 of IMP
   • Have students read *A Year 2 Sampler*
   • Review grading policies, expectations, and so on
3. Have an "icebreaker" introduction activity

4. Reflect on Year 1

### At Home

Homework 1: *Math, Me, and the Future*

### Special Materials Needed

• Students' portfolios from Year 1

## 1. Forming Groups

At the beginning of the unit, students should be formed into new groups as described in the IMP *Teaching Handbook*. We recommend that new groups be formed again on Day 11 and on Day 21.

## 2. Overview of the Year: *A Year 2 Sampler*

The first day of this unit has a sequence of brief activities designed to welcome students to a new year of study, to have them review some of what made Year 1 of IMP special, and to get them to think about goals for the coming year.

You can introduce Year 2 of IMP by giving a short summary of each of the units that students will study this year. *A Year 2 Sampler* provides students with such a summary in written form.

• *Grading policy, expectations, and so on*

You will probably want to take some time to discuss your grading policy and general expectations of classroom behavior with the class.

*"Why do you think it's important to do homework regularly?"*

*"What can you do if you don't understand an assignment?"*

In particular, this is a good time to emphasize the importance of doing homework regularly. Even if students can't do an assignment completely, they should at least make a start on it. Remind them that they can use each other as resources and that if for some reason they can't do a given assignment, they should write about *why* they can't do it.

• *Materials*

Go over with students the list of materials that they will need to supply. For this unit students should provide

- a notebook or binder

- a pencil

- lined paper

- graph paper

- a 12-inch ruler with both inch and centimeter scales

- an assortment of colored pencils or pens

- a protractor

- a calculator

Students need not bring their calculators to class, because graphing calculators will be provided. But they should have calculators available every night for use in the homework.

# 3. Getting to Know You

The initial activity suggested for today is an "icebreaker" whose purpose is to get students talking again and sharing ideas, facts, thoughts, and so on. This activity should help to reestablish the IMP camaraderie that we hope students felt by the end of last year.

Tell students that they will have exactly one minute each to talk about themselves within their groups. They can talk about last year, about their summer, about their feelings about mathematics, about their families—anything they want—but they will each have their groups' attention, uninterrupted, for 60 seconds. After one minute, you will tell the first student to stop and the next to begin, and so on.

You may want to emphasize the importance of having the other three members of the group be good listeners.

To avoid having students waste time deciding who should go first, you can give them this order: clubs, diamonds, hearts, spades.

When all students have had their turn, you can go directly to the next activity.

# 4. Reflections on Year 1

Tell students that the next activity involves reflecting on their first year of IMP and that tomorrow they will look at some problems involving situations from Year 1 units.

## • *Reflecting on how IMP is different*

As the first stage of this reflecting, ask students to do focused free-writing about how Year 1 of IMP differed from other mathematics courses that they have taken or that friends are presently taking.

You may need to review the idea of focused free-writing with students (see the IMP *Teaching Handbook*). These are some key points.

- Focused free-writing is not to be collected, although students should have the opportunity to read some of it out loud or just share their ideas.

- Students should write, write, write—a sort of stream of consciousness.

- Students should try to focus on the assigned topic.

*"What are some of your ideas about how IMP was different?"*

After students wrap up their writing, lead a class discussion on their ideas. You may want to jot down their thoughts on chart paper.

## • *Reflecting on Year 1 units*

In the next stage of reflection, students will think back to the specific units from Year 1. One way to set this up is to have students decide on their favorite unit and then have students with the same favorite meet together. (For example, designate a corner or area of the room for each unit.)

> *Reminder:* The IMP units from Year 1 are
>
> - *Patterns*                                    - *The Pit and the Pendulum*
>
> - *The Game of Pig*                          - *Shadows*
>
> - *The Overland Trail*

Students in the same group should then spend a few minutes discussing what important mathematics they studied in that unit. Then someone from each group should give a brief report to the class.

## • *Reflecting on portfolios*

As the last stage of this reflection process, return students' mathematics portfolios from last year and have them look over the portfolios.

These portfolios should contain students' cover letters on each of the five units, samples of their work from each unit, and the end-of-unit and semester assessments for the year.

We intend this portfolio examination not as a review of the mathematics of the previous year, but more as an opportunity for students to get a sense of what their mathematical writing was like. They may enjoy seeing how they changed over the year.

You can suggest to students that they focus on the longer items in their portfolios, perhaps looking first at their cover letters for each unit and then at their POWs.

Encourage students to do this portfolio examination with their group members, sharing items and perhaps using the portfolios to jog their memories about what math class was like. If the class includes students who did not use IMP last year, this discussion can introduce them to some of what the rest of the class did.

Tell students that as they share in their groups, they should be thinking and talking about expectations for themselves for this coming year.

## • *Using portfolios in the future*

Tell students that they will probably find it helpful to use their portfolios to review ideas from Year 1 that come up again in Year 2 (and later). For example, they may find material about standard deviation and normal distributions (from *The Pit and the Pendulum*) very helpful when they get to *Is There Really a Difference?*

### • *For teachers: What to do with the portfolios in the future?*

Much discussion has taken place around this question, and teacher groups do not agree on the answer. While we recommended for Year 1 that teachers hold onto students' portfolios, many teachers think that by Year 2 students should become the keepers of their own collection of unit portfolios.

If you collect the portfolios, then students cannot lose them, and they will always be available in class for students to use. You will also have them available for discussions with students about their progress or for parent conferences.

On the other hand, if you let students keep their portfolios, then they will be able to use them at home and can bring them to school if needed. The responsibility for keeping them will then belong to the students. Of course, a potential risk is that you will never see the portfolios again.

One other important consideration is whether your school has enough room to store four years' worth of portfolios and whether you are willing to transfer them from one class to the next.

## *Homework 1: Math, Me, and the Future*

In this opening assignment, students are asked to reflect on their first year in IMP and their preparedness for it.

They also write letters to themselves to help them set goals and motivate themselves for the coming year.

You may want to tell students that you will not read the letters that students write to themselves. There is a reminder at the end of *Is There Really a Difference?* that these letters should be delivered back to students.

***Chris Quinn thinks back on last year's units as he works through "Memories of Yesteryear."***

# A Year 2 Sampler

You are now beginning Year 2 of the Interactive Mathematics Program. As with Year 1, the curriculum is organized around major units. Here's a brief look at each of these units.

## Solve It!

The first unit of Year 2 focuses on the use of equations to represent problems and on techniques for solving equations, including the basics of algebraic manipulations.

## Is There Really a Difference?

In this unit you'll look at how you might decide whether differences that show up in samples from two populations necessarily represent real differences in the overall populations. You will be introduced to important statistical concepts such as the *null hypothesis* and the *chi-square distribution*.

## Do Bees Build It Best?

In this unit you'll look at the geometry of bees' honeycombs and ask whether this shape is the most efficient. You will work with fundamental ideas about area and volume, learn about the Pythagorean theorem, and continue your work with trigonometry.

*Continued on next page*

## Cookies

The central problem of this unit involves a bakery's decision about how many of each of two kinds of cookies to make. The owners have certain restrictions on oven space, baking time, and so on, and they want to allocate resources in a way that will maximize their profit. You will solve their problem by graphing linear equations and inequalities, solving systems of linear equations with two variables, and reasoning with graphs.

## All About Alice

This unit uses a metaphor from *Alice in Wonderland* to help you explore concepts about exponents and logarithms. It includes material about scientific notation, order of magnitude, and significant digits.

# Homework 1    Math, Me, and the Future

These three writing assignments are designed to help you set the tone for your work in mathematics for this year.

1. Write about how your first year in IMP affected the way you think about mathematics.

2. Write about how the mathematics classes you had before IMP prepared you for Year 1 of IMP.

3. On a separate sheet of paper, write a letter to yourself. Set goals for yourself for this year, imagining that you are your own conscience. You can ask yourself some questions and give yourself some reminders about what you would like to do to succeed in mathematics this year.

Tomorrow in class you will put this letter in an envelope that you address to yourself. Your letter will be delivered to you later in the year.

# DAY 2 Memories of Yesteryear

## Mathematical Topics

- Representing problems encountered in Year 1 of IMP symbolically

## Outline of the Day

### In Class

1. Discuss *Homework 1: Math, Me, and the Future*

   - Have students address envelopes to themselves and place their letters inside

2. Introduce the new unit

   - The unit has no central problem
   - The key ideas of the unit are
     ✓ representing problems by equations
     ✓ solving equations in various ways

3. *Memories of Yesteryear*

   - Students represent simple situations using equations
   - The activity will be discussed on Day 3

### At Home

*Homework 2: Building a Foundation*

## Discuss With Your Colleagues

### Do They Remember Anything?

The first mathematical activity of the unit, *Memories of Yesteryear*, involves situations and ideas from Year 1. Sometimes teachers are disappointed in students' retention level from the end of one school year to the beginning of another. Over the next few days, as students work with ideas from Year 1,

observe how retention in IMP students compares with that of students in a
traditional program, and discuss your observations with your colleagues.
What's the best way to review material that you think students should
already know?

## 1. Discussion of *Homework 1: Math, Me, and the Future*

You can have students share ideas in their groups or as a whole class
concerning parts 1 and 2 of the homework.

Hand out blank envelopes and have students self-address the envelopes and
enclose their letters to themselves (from part 3 of the assignment). You can
remind students that they will get these back later in the year. (There is a
reminder to you at the end of *Is There Really a Difference?*)

## 2. Introduction to *Solve It!*

Before turning to the first mathematical activity of the unit, take some time
to introduce the unit. Tell students that unlike most IMP units, this one does
not have a central problem to solve. Instead, it introduces some important
ideas from algebra that are useful in a wide variety of situations. Inform
students that they will be revisiting contexts and concepts from the units
from Year 1 to see how the ideas from algebra can be applied.

In particular, you can tell students that the unit will focus on two major
ideas:

- The use of equations and algebraic notation to express problem
  situations

- The development of a variety of techniques for solving equations,
  including guess-and-check, graphing, and algebraic manipulations

*Note:* The guess-and-check method is sometimes also called "trial and error."
We will use the two terms interchangeably and we recommend that you let
students know that they are synonymous.

---

### • *Building on Year 1*

Many of the problems in this unit
(including those in today's activity,
*Memories of Yesteryear*) are adapted
from situations discussed in units
from Year 1. The purpose of using
such problems is

- to review ideas from those units

- to show the variety of situations
  in which equations arise

- to give students a chance to solve
  equations in familiar and meaningful contexts

---

In addition to the problems in today's activity, the unit includes these problems that build on work from Year 1.

- Question 3 of *Homework 4: Running on the Overland Trail,* which uses ideas about rate of consumption from *The Overland Trail*

- *Lamppost Shadows* (Day 5), which uses ideas about similarity from *Shadows*

- *Catching Up* (Day 9), which uses ideas about rate of travel from *The Overland Trail*

- *Homework 28: Swinging Pendulum,* which involves the formula for period in *The Pit and the Pendulum*

Some of these problems will be reexamined in *Old Friends and New Friends* on Day 22.

## 3. Memories of Yesteryear

Introduce this activity to the class as a starting point for the exploration of solving equations.

### • *Why use equations?*

It is very important that you acknowledge to students that these problems can be answered without using equations. Many students may think it's pointless to bother with equations or algebra for some of the problems in this activity—and for other problems later in the unit as well.

You may need to point out to students several times in the early days of this unit that they are being asked to define variables and write equations for these simple problems so that they will be comfortable with these tools when they encounter more difficult situations later.

### • *Getting started on the activity*

Have groups begin work on the activity *Memories of Yesteryear.* The focus of their work should be on defining variables, setting up equations, and then solving the equations (probably by a guess-and-check approach).

As groups finish, you can give them an overhead transparency and pen and assign one of the problems for them to present to the class tomorrow. You may want to assign the problems in reverse order of difficulty so that the last group to be assigned a problem will not have to do a long one.

*"What is the problem asking you to find?"*

If you find that students are having trouble getting started on a particular question, you might give them some help with defining the variables for that problem. One strategy is "let the variable answer the question." With this method, students define a variable so that when they identify its value, they have an answer to the question. For example, in Question 1, the question is, "How many cubes did the chef throw in during the middle handfuls?" So a natural choice would be to let the variable represent this number of cubes.

This is an obvious approach for some students, and if groups are not having difficulty defining variables, you need not mention it. However, "letting the variable answer the question" might be helpful for students or groups who are having trouble getting started.

Another useful strategy for some problems, especially ones involving geometry, is to make a sketch and simply label every unknown length with a variable. Students can then write any true equations that fit the situation until they find one whose solution will help to answer the question. (This method can lead to an overabundance of variables, but it still might be helpful for groups having difficulty getting started.)

*Note:* Question 1 refers indirectly to the hot-and-cold-cube model for positive and negative numbers that students used in *Patterns*. But the question does not involve negative numbers, and there is no need to go over the model at this time. Students will first need to work with negative numbers in this unit when they get to *Homework 5: 1-2-3-4 Puzzle with Negatives*. We suggest that you wait to see how students do with that assignment and review the hot-and-cold-cube model as needed in the context of that homework discussion.

Students will have several further opportunities to strengthen their understanding of the arithmetic of integers, beginning with the review of substitution on Day 8.

## Homework 2: Building a Foundation

Question 1 of tonight's homework continues the theme of writing and solving equations. But the main mathematical focus is the principle that the area of a rectangle can be found by multiplying its length by its width.

The concept of area used here is an intuitive one, based on counting tiles.

Students will use this concept later in the unit, still on an intuitive level, in the development of the distributive property. The problems in this assignment lay some groundwork for that development. Area will be studied more formally in the Year 2 unit *Do Bees Build It Best?*

# Memories of Yesteryear

In this assignment, you will be solving problems based on situations that may be familiar to you. Although you might prefer to solve some of the problems without equations, your assignment is to use variables and equations according to these steps.

- Choose the variable you are going to use in each problem and state what it represents.

- Write an equation, using your variable, that represents the problem.

- Solve the equation and the problem using any method you wish, including guess and check (also known as "trial and error").

1. From *Patterns*

   A chef put several batches of cubes into a cauldron. The first batch contained 27 cubes. The last batch contained 56 cubes. A total of 108 cubes were put into the cauldron.

   How many cubes did the chef throw in that were not part of the first or last batch?

2. From *The Overland Trail*

   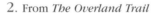

   Each adult needed 5 yards of shoelace for the trip to California and each child needed 3 yards. A certain family with seven children needed 71 yards of shoelace.

   How many adults were in the family?

*Continued on next page*

Interactive Mathematics Program                                              7

Al wins a
fixed number of points

Betty wins 2 points

**3.** From *The Game of Pig*

Al and Betty were playing a game using the spinner shown here. Betty won 2 points from Al every time the spinner landed on the white section.

Al won a fixed amount from Betty when the spinner landed in the gray section, but you don't know what that amount was. You do know that after 100 games, the results matched the probabilities perfectly, and Al was 25 points ahead overall.

How much did Al win from Betty each time the spinner landed in the gray section?

**4.** From *The Pit and the Pendulum*

A group was doing experiments in order to find the period of a pendulum in terms of its length. The group came up with this rule:

If you take the square root of the length of the pendulum (in inches) and multiply that by 0.32, then you will get the number of seconds for one period of the pendulum.

If the period of a certain pendulum is 3.84 seconds, then what is the pendulum's length (based on this rule)?

**5.** From *Shadows*

A person who is 6 feet tall is standing 3 feet from a small mirror that is lying flat on the ground. By looking in the mirror, the person can see the top of a tree that is 15 feet from the mirror.

How tall is the tree?

# Homework 2     Building a Foundation

1. Maisha is planning to build a patio along the back wall of her house, which is 32 feet long. The patio will be rectangular in shape and will fit against the full length of the back wall (so one side of the patio will be 32 feet long).

   The patio will be built out of square tiles that are 1-foot-by-1-foot. Maisha is thinking about this question:

   > If I have 256 tiles to work with, how far out from the wall will the patio extend?

   Pretend you are Maisha and do these tasks:

   - Make a sketch of the situation.
   - Choose the variable you are going to use and state what it represents.
   - Write an equation that represents the problem.
   - Solve the equation and the problem using any method you wish (including trial and error).

2. Benito is also going to build a patio, but his patio does not have to fit exactly against a wall. In fact, all that Benito has decided is that the patio should be rectangular in shape and should use all of the 144 tiles he has available. (Like Maisha, he is using square tiles that are 1-foot-by-1-foot.)

   Find as many possibilities as you can for the dimensions of Benito's patio. (*Note:* You do not need to go through all the steps you used in Question 1.)

# DAY 3

# *Memories of Yesteryear, Continued*

## Mathematical Topics

- Understanding the area of rectangles
- Defining variables
- Writing algebraic representations of problems
- Solving equations using a guess-and-check method

## Outline of the Day

### In Class

1. Discuss *Homework 2: Building a Foundation*
   - Establish the connection between the area of rectangles and multiplication

2. Discuss *Memories of Yesteryear* (from Day 2)

3. Introduce *Homework 3: You're the Storyteller*
   - Have students make up some problems to go with a simple equation

### At Home

*Homework 3: You're the Storyteller*

*Note:* You may want to use the first five to ten minutes of class to have each group prepare a presentation for either last night's homework or yesterday's in-class activity.

# 1. Discussion of Homework 2: Building a Foundation

*Comment:* You may want to use the beginning of *Solve It!* to emphasize the importance of doing homework regularly. Today, you might discuss different ways that students found to get started on the assignment and remind students about the importance of at least making an effort on every homework assignment.

## • Question 1

*"What exactly does t represent?"*

In Question 1, pay close attention to how students define the variable they use. For instance, if a student says, "*t* is the number of tiles," you can bring out the ambiguity of this statement by asking if it means the total number of tiles, the number of tiles along the wall of the house, or the number of tiles out from the wall.

As noted previously, students may rebel against bothering with a variable and an equation in a problem like this, because they can probably solve the problem without a formal equation. But you should insist that they write something like $32t = 256$, or even $t = 256 \div 32$, simply for practice. Remind them that they will be doing harder problems for which the naming of variables and writing of equations will be more important.

## • Multiplication, tiles, and area

*Reminder:* Students will use the concept of area later in this unit as part of the development of the distributive property, so at this time it is important to discuss the relationship between multiplication and the number of tiles.

Use the discussion of Question 1 to make a transition between "number of tiles" and area. That is, bring out to students that in counting the number of tiles, they are also finding the area of the rectangular patio (measured in square feet, because each tile is one foot square).

*"Why are you using multiplication? What does multiplication have to do with counting tiles?"*

You can begin this transition by asking students why they are using multiplication in their equations or what multiplication has to do with the process of counting tiles.

*"How could you know that there are 12 tiles without counting them?"*

If they need help, you can suggest that they focus on a specific, simple example, such as a 3-tile–by–4-tile arrangement. Ask the students how they could know that there are 12 tiles without counting them explicitly. (The goal here is to make a connection between the repeated addition of a given number—which defines multiplication—and the more visual representation involving rows with an equal number of tiles).

Students should see that there are two possibilities. They can have 3 rows with 4 squares in each, as shown here.

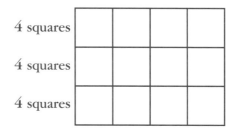

4 squares

4 squares

4 squares

Or they can have 4 rows with 3 squares in each, again as shown here.

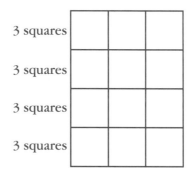

3 squares

3 squares

3 squares

3 squares

Both cases involve addition—either three 4's added together or four 3's added together. Thus, the multiplication problem 3 · 4 means either 4 + 4 + 4 or 3 + 3 + 3 + 3, and can be visualized as either three rows of four squares each or four rows of three squares each.

Help students to see that any product of two whole-number factors can be thought of in terms of a rectangular area.

You may want to point out that the connections discussed here—between dimensions, area, and multiplication—are implicit in such usage as labeling a rug that is 9 feet wide and 12 feet long as "9 × 12," using the multiplication sign instead of the word "by."

## • *Question 2*

*"What are some of Benito's choices?"*

On Question 2, you can have the class list the possible choices of dimensions for Benito's patio. Students should see that this is the same as identifying how many ways 144 can be written as a product of two whole numbers. (They may disagree about whether a 6 × 24 patio is the same shape as a 24 × 6 patio. Let them resolve this any way they like, or leave it unresolved.)

*Note:* In *POW 3: Divisor Counting* (assigned on Day 17), students will explore what it is about a positive integer that determines how many whole-number divisors it has. Question 2 from last night's homework is, in part, an anticipation of that activity.

## 2. Discussion of *Memories of Yesteryear*

*"What other ways did you have for solving the problem?"*

As each group presents its problem, encourage other groups to share alternative ways to express the problem. Once again, emphasize the need for clear definitions of the variables used.

After the discussion, you may want to collect this assignment to assess the skill level of the class as a whole and to see if any students are having particular trouble.

### • Questions 1 and 2

The first two problems are the simplest ones and are probably the ones for which students are most likely to see variables and equations as irrelevant. You may need to assure them that these problems need nothing more complicated than equations such as $27 + c + 56 = 108$ (where $c$ is the number of cubes in the middle handfuls) and $a \cdot 5 + 7 \cdot 3 = 71$ (where $a$ is the number of adults in the family).

*Comment:* Although students might rewrite $a \cdot 5$ as $5a$, this is not essential. (The use of the notation $5a$ as shorthand for multiplication of $a$ by 5 was discussed in *Patterns.*) It is more important that the notation express students' natural understanding of the arithmetic than that they use the most standard notation. If there were, say, 9 adults, many students would calculate the amount of shoelace as $9 \cdot 5$, rather than as $5 \cdot 9$, so for them, $a \cdot 5$ would be more natural. You can mention that the multiplication dot is normally used when the variable comes first in the product (as in $a \cdot 5$) but is omitted when the coefficient comes first (as in $5a$).

### • Question 3

*"How did your answer depend on the specific shading of the spinner?"*

For Question 3, have presenters articulate how they used the specifics of the spinner to analyze the problem. They will probably say something like, "The white section is $\frac{3}{4}$ of the area, so Betty should win about $\frac{3}{4}$ of the time, which would be 75 games. That leaves 25 games for Al to win." This would lead to an equation like $25g - 75 \cdot 2 = 25$ (where $g$ represents what Al wins from Betty if the spinner lands in the gray area).

### • Questions 4 and 5

Students' main task for Question 4 is to put the verbal description of the rule into an algebraic form. The equation might be $3.84 = 0.32 \sqrt{L}$ (where $L$ is the length of the pendulum).

*"What diagram did you use for Question 5?"*

You can use Question 5 as an occasion to review some of the ideas about similarity from *Shadows.* Be sure to get a diagram of this problem from the presenters.

As with most similar-triangle problems, there are several ways to set up an equation of proportionality for this situation. The equations $\frac{x}{15} = \frac{6}{3}$ and $\frac{6}{x} = \frac{3}{15}$ are just two of the possibilities.

## 3. Introduction to *Homework 3: You're the Storyteller*

*"What situation might lead to the equation 2x = 8?"*

In *Homework 3: You're the Storyteller*, students are asked to reverse the usual procedure of writing an equation to fit a situation. Here they are given the equations and are asked to make up situations to fit them.

Because this is an unusual task, students will probably benefit from spending a few minutes in class looking at a simple example. For instance, ask them to come up with some possible situations that would lead to the equation $2x = 8$.

If needed, offer some examples done in the format you desire, such as

- *Problem:* Donna studied for twice as long as Jerod. Donna studied for 8 hours. For how long did Jerod study?

  *Variable:* Let $x$ represent the number of hours Jerod studied.

- *Problem:* Sven is twice as old as his sister Tanda. Sven is 8 years old. How old is Tanda?

  *Variable:* Let $x$ represent Tanda's age.

## Homework 3: You're the Storyteller

Students will see more complex examples of this task later in the unit.

# Homework 3　　　　You're the Storyteller

In *Memories of Yesteryear,* you started from situations and created equations to fit those situations. In this assignment, you will work in the opposite direction, creating situations that fit the five equations given here. This task has three steps.

• Create a situation.

• Write a question about the situation so that solving the equation will give you the answer to your question. State clearly what the variable in the equation represents in the situation.

• Solve the equation to answer your question.

1. $4a = 12$

2. $r + 5 = 20$

3. $2m + 1 = 11$

4. $\frac{t}{3} = 8$

5. $13 - f = 6$

# DAY 4

*Is It a Digit?*

## Mathematical Topics

- Creating questions that can be solved by a given equation
- Defining variables
- Working with proof

## Outline of the Day

### In Class

1. Discuss *Homework 3: You're the Storyteller*

2. *Is It a Digit?*
   - Students use trial and error to solve a puzzle (in preparation for their new POW)

3. Discuss *Is It a Digit?*
   - Be sure students see what makes a proposed solution correct or incorrect

4. Introduce *POW 1: A Digital Proof*
   - Emphasize that students need to *prove* that they have all possible solutions

5. Discuss the new *Standard POW Write-up*

### At Home

Homework 4: *Running on the Overland Trail*

*POW 1: A Digital Proof* (due Day 11)

## 1. Discussion of *Homework 3: You're the Storyteller*

You can assign an equation to each group and have group members share their work for that equation. The group should prepare a presentation on a problem situation that one of the members created for that equation. You can have spade card students make the presentations. You might also have each group choose one favorite example to share, even if it's not necessarily from the assigned equation.

Many students have trouble getting started on problems such as these, so you might have the creator of each problem share with the class how she or he got the idea for the problem.

## 2. *Is It a Digit?*

This problem is a preliminary stage of *POW 1: A Digital Proof.* A solution can be found through trial and error, but the task of showing that no other solutions exist forces students to think systematically. In *Is It a Digit?* students will find a solution. Their task in the POW will be to prove uniqueness.

Begin by pointing out to the class that this activity is introductory to the POW and not part of the main development of the unit. Then have a student read the problem to the class.

*"What must happen if you put a 3 in the box labeled '4'?"*

Have a volunteer explain what the significance is, for example, of putting a 3 in the box labeled "4." Be sure students understand that this means that there must be exactly three 4's used altogether in the boxes.

Next, have groups begin work on finding a solution. Avoid giving more examples than the one just stated. Have students wrestle with the idea with their fellow group members.

As groups begin to come up with a solution, give one group an overhead transparency and pen to prepare a presentation of its work. Tell group members that their presentation should consist solely of a statement of what their solution is and an explanation of why it fits the conditions of the problem.

## 3. Discussion of *Is It a Digit?*

*Reminder:* This activity is intended only as an introduction to the POW. Have students discuss the solution, but do not get into the discussion of why it is unique.

When the presenting group is ready and the other groups have made some progress—they should at least know what they are looking for—have the presenting group give its solution. As the accompanying diagram shows, the presenter should point out that the box labeled "0" has a 2 in it, which is matched by the fact that there are two 0's used (in the boxes labeled "3" and "4"). A similar explanation should be given for each of the five boxes.

| 2 | 1 | 2 | 0 | 0 |
|---|---|---|---|---|
| 0 | 1 | 2 | 3 | 4 |

*"Did anyone find
any other
solutions?"*

Ask if anyone found any other solutions. If students think they have other solutions, have the class look at what they propose and find out what's wrong with it.

*Note:* It's okay if students recognize that there are no other solutions. That will help them focus on the fact that the main task in the POW is to *prove* uniqueness.

# 4. Introduction to
## *POW 1:*
## *A Digital Proof*

After presentation of a solution to *Is It a Digit?* have students read the POW. Emphasize that their task is to provide a completely convincing argument that there are no solutions other than the one they find. You might point out that in this problem, a completely convincing argument probably involves a careful, case-by-case analysis. You may want to review *Homework 2: Who's Who?* from the Year 1 unit *Patterns* as the basis for some discussion of this approach to proof.

This POW is scheduled for discussion on Day 11.

### • *How to eliminate cases*

If students need some help getting started, you can suggest that one way to proceed is to eliminate cases until all that's left are the known solutions. You also might give them a single example of this process. However, it's a good idea to refrain from any discussion with the class beyond that example lest you do too much for the students and not let them figure things out on their own.

As an example of this "case-by-case elimination" method, you can ask if anyone can explain why there cannot be a 3 in the box labeled "4." (If necessary, ask what would have to happen if there were a 3 in that box.) One response might be that if there were a 3 in that box, then three of the other boxes would have 4's in them, which would leave only one open box. There would be no room for four of each of the numbers that have a 4 in their box. (If needed, you can pick a specific choice for placement of the three 4's to illustrate.)

Students often refer to this type of reasoning as a "chain reaction": If *this* were true, then *this* would be true, and then *this* would be true, and so on.

# 5. For Reference:
## *The Standard POW*
## *Write-up*

Students should generally be quite familiar with what is expected in POW write-ups, as detailed in *The Standard POW Write-up*. However, point out the following two changes from the Year 1 version of the write-up.

- The category of *Extensions* has been removed, because the POWs for Year 2 are already stated in fairly general terms.

- A new category of *Self-assessment* asks students to assign themselves a grade for their work on the POW and then justify that grade.

You should take some time to discuss what role, if any, the self-grading that students do for this new category will play in the grades that you assign them. This is also a good time to present your general expectations for the POWs.

### • *Optional: Having students create a rubric*

If your students have some experience with the use of rubrics for evaluating written work, or if you wish to introduce this idea to them, you may want to have them create rubrics to use in grading POWs. One method is to assign this task to a group at the same time that the POW itself is assigned. Emphasize that the rubric the group creates should go beyond general principles—that is, it should be specific to the particular POW.

This will probably work best if you can allow group members some time to work together on both the POW itself and the rubric. If this is not feasible, have individual group members come up with rubrics and then take some class time to synthesize their work. (This can be done either with just you and that group or with the whole class.)

Ideally, the group will complete its rubric a couple of days before the POW write-ups are due. This will enable you to discuss it with group members and give them time to make needed changes. The group can then present the rubric to the rest of the class before the students finish their write-ups.

You will then have the option of letting students grade one another's POW write-ups based on that rubric, either instead of or in addition to your own grading. If the rubric proves problematic for certain write-ups, you can discuss this with the class.

If you adopt this process, keep in mind that students' early efforts may not be very good. Think of the process of having students create rubrics as a learning experience for them, and look for growth over time.

*Note:* As an alternative to complete adoption of this idea, you might either do it only occasionally or use it as an honors project.

## Homework 4: Running on the Overland Trail

This assignment continues the theme of using variables and equations to solve problems. Students might consider the first two questions to be too easy to require the use of algebra, but the third is a bit more complicated.

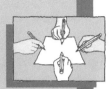

# *Is It a Digit?*

There are five empty boxes shown here labeled 0 through 4.

Your task is to put a digit from 0 through 4 *inside* each of the boxes so that certain conditions hold:

- The digit you put in the box labeled "0" must be the same as the number of 0's you use.

- The digit you put in the box labeled "1" must be the same as the number of 1's you use.

- The digit you put in the box labeled "2" must be the same as the number of 2's you use, and so on.

Of course, you are allowed to use the same digit more than once.

You may want to make several copies of the set of boxes in order to try various combinations of digits.

## *What Not to Do*

Here is an example of an *incorrect* way to fill in the boxes.

| 2 | 3 | 1 | 2 | 2 |
|---|---|---|---|---|
| 0 | 1 | 2 | 3 | 4 |

This is incorrect for many reasons. For instance, there is a 1 in the box labeled "2," but there is more than one 2 used in the boxes. Similarly, there is a 2 in the box labeled "4," but the number of 4's used is not equal to 2.

# POW 1        *A Digital Proof*

☐     ☐     ☐     ☐     ☐

0       1       2       3       4

In *Is It a Digit?* you looked for a way to fill in the numbered boxes shown here in a way that fit certain conditions. Your task in this POW is to *prove* that you have all the solutions. (If you haven't yet found a solution, then doing so is also part of your POW.)

## Write-up

1. *Problem Statement:* Explain the problem from *Is It a Digit?*

2. *Process:* Based on your notes, describe how you went about finding all of the solutions to *Is It a Digit?* and how you decided that you had them all.

3. *Solutions:* List all solutions you found for *Is It a Digit?* Then write a careful and detailed proof that there are no solutions to *Is It a Digit?* other than those you listed.

4. *Evaluation*

5. *Self-assessment*

(For write-up categories with no specific instructions, use the description in *The Standard POW Write-up.*)

# The Standard POW Write-up

Each POW is unique, and so the form of the write-up may vary from one POW to the next. Nevertheless, most of the categories that you will be using for your POW write-ups will be the same throughout the year.

The list below gives a summary of the standard categories for Year 2.

Some POW write-ups will use other categories or require more specific information within a particular category in order to make the write-up more suitable to the POW. But if the write-up instructions for a given POW simply list a category by name, you should use the descriptions below.

*Continued on next page*

# *The Standard POW Write-up Categories*

1. *Problem Statement:* State the problem in your own words. Your problem statement should be clear enough that someone unfamiliar with the problem could understand what it is that you are being asked to do.

2. *Process:* Describe what steps you took in attempting to solve this problem, using your notes to jog your memory. Include steps that didn't work out or that seemed like a waste of time. Complete this part of the write-up even if you didn't solve the problem. And if you got help of any kind on the problem, indicate what form it took and how it helped you.

3. *Solution:* State your solution as clearly as possible. Explain why you think your solution is correct and complete. (If you obtained only a partial solution, give that. If you were able to obtain more general results, include them.)

   Your explanation should be written in a way that will be convincing to someone else—even someone who initially disagrees with your answer.

4. *Evaluation:* Discuss your personal reaction to this problem. For example, you might comment on these questions.

   • Did you consider the problem educationally worthwhile? What did you learn from it?

   • How would you change the problem to improve it?

   • Did you enjoy working on the problem?

   • Was the problem too hard or too easy?

5. *Self-assessment:* Assign yourself a grade for your work on this POW, and explain why you think you deserve that grade.

# Homework 4

# Running on the Overland Trail

For each of the problems here, complete these tasks.

- Choose a variable.

- State clearly what the variable represents.

- Write an equation using your variable that represents the problem.

- Solve both the equation and the problem.

1. If Phillipe had $7 more, he could buy a $30 pair of tennis shoes. How much money does he have?

2. Yolanda jogged 2 miles to a lake, ran twice around the lake, and then jogged 2 more miles home. Altogether she traveled 10 miles. How far is it around the lake?

3. An Overland Trail family is carrying 5 gallons of water per person in its wagon. Unexpectedly, two stragglers join the group. The family figures out that this means there are now only 4 gallons per person. How many people were in this Overland Trail family?
(*Hint:* Take a guess and write down what you should do to see if it's right. Keep doing this until you see a pattern in the arithmetic steps. Then use these steps to come up with an equation.)

Interactive Mathematics Program     15

## Mathematical Topics

- Representing problem situations by equations
- Using similarity to develop equations

# Outline of the Day

## In Class

1. Discuss *Homework 4: Running on the Overland Trail*

   - Describe the equations as the **algebraic representations** of the problems

2. *Lamppost Shadows*

   - Students use equations to represent problems involving similar triangles
   - The activity will be discussed on Day 6

## At Home

*Homework 5: 1-2-3-4 Puzzle with Negatives*

## 1. Discussion of *Homework 4: Running on the Overland Trail*

You may want to ask for volunteer presentations on these problems.

If students had difficulty with Question 3, discuss how to use the guess-and-check approach suggested in the problem. The key step is using the number of people in the family to figure out the total amount of water.

For example, suppose one guess is that the family has 15 people. This means that they had 15 • 5 (or 75) gallons of water. With the two additional people, they would have $\frac{75}{17}$ (about 4.4) gallons per person. A useful question for some students is, "How does this tell you that 15 isn't the right answer?"

Using an In-Out table may help some students to develop an expression such as $\frac{5n}{n+2}$ that represents how much water there will be per person if there are $n$ people in the family. You can use this example to review the language of functions, using a statement such as, "The amount of water per person is a function of the number of people in the family."

There are several different equations that students might use for this problem, including $\frac{5n}{n+2} = 4$ and $n \cdot 5 = (n + 2) \cdot 4$. The first of these focuses on the amount of water per person while the second looks at the total amount of water in two ways.

## • Overall goals

You should help students to see these ideas.

- Each problem can be described by an equation.

- Each of these three equations has a solution.

- The solutions to the equations are also the answers to the problems.

Here are some possibilities for the equations.

- For Question 1: $p + 7 = 30$ (where $p$ is the amount of money Phillipe has)

- For Question 2: $2 + 2d + 2 = 10$ or $2d + 4 = 10$ (where $d$ is the distance around the lake)

- For Question 3: $\frac{5n}{n+2} = 4$ (where $n$ is the number of people in the family)

*"What does it mean to 'solve an equation'?"*

In discussing these examples, be sure to use the phrase "solving the equation." Ask students to state what this means. They will probably come up with a statement like "finding the number that can be substituted for the variable to give a true statement." Refer to this number as a **solution** to the equation—that is, a value for the variable that makes it true—and to the equation as the **algebraic representation** of the problem. Emphasize that solving an equation means finding *all* the solutions.

*Comment:* Because the equations in these problems each have only one solution, you may want to introduce another simple example, such as the equation $x^2 = 9$, to bring out the last point.

Students may want to write equations for Questions 1 and 2 that actually *solve* the problems. For instance, they might write $p = 30 - 7$ as the equation for Question 1 and $d = \frac{10 - 2}{4}$ as the equation for Question 2. Encourage them, however, to create equations in which the variable is embedded in the equation, because such equations reflect the sense of the situation as described and also suggest the most obvious way to check a guess.

You may find it helpful here to suggest that students view the "result" in these two problems ($30 cost for the shoes or 10 miles total jogged) as similar to the *Out* of an In-Out table—that is, as something that depends on the value chosen for $p$ or $d$.

## 2. Lamppost Shadows

The next activity, *Lamppost Shadows*, uses the context of similarity to look at the use of equations and variables. Because students probably will be unable to solve Question 2 by trial and error, this problem will help motivate the use of algebraic methods.

This activity will be discussed tomorrow.

Encourage students to use informal methods for solving the equations in this activity. Try to keep them from getting bogged down looking for an exact answer to Question 2c.

If students need help with Question 1b, you might suggest that they make a diagram showing two separate similar triangles. Labeling the vertices might also be helpful to some students. If the issue comes up, acknowledge that other proportions could be used for this situation.

## Homework 5: 1-2-3-4 Puzzle with Negatives

This problem is the same as *POW 2: 1-2-3-4 Puzzle* in *Patterns* except that it calls for the use of negative numbers. The goal here is to begin a review of negative-number arithmetic and of the conventions for order of operations.

# Lamppost Shadows

Chantelle and Nelson are members of a volunteer clean-up committee. At the end of the day they are waiting with other volunteers for the shuttle back to the community center.

1. Chantelle is 5 feet tall. She is standing 30 feet from a lamppost that is 25 feet tall. Using $S$ to stand for the length of Chantelle's shadow, you can represent her situation using the diagram shown here.

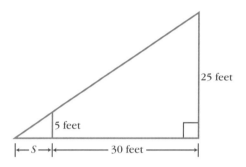

a. Take a guess as to how long the shadow is, *just from looking at the diagram*.

b. Explain, using the diagram, why $S$ must fit the equation

$$\frac{S}{S + 30} = \frac{5}{25}$$

c. Try to find a number for $S$ that solves this equation. If you can't solve the equation exactly, look for a number that comes close.

d. Compare your answer in Question 1c to your guess in Question 1a. Does your answer in Question 1c seem reasonable?

2. Nelson, who is 6 feet tall, is standing 20 feet from the same lamppost.

a. Draw and label a diagram showing Nelson and his shadow.

b. Write an equation whose solution would give the length of Nelson's shadow.

c. Try to find a number that solves your equation from Question 2b. If you can't solve the equation exactly, look for a number that comes close.

# Homework 5

# 1-2-3-4 Puzzle with Negatives

This assignment is based on *POW 2: 1-2-3-4 Puzzle* from the Year 1 unit *Patterns*. The idea of that problem was to use the digits 1, 2, 3, and 4 once each, along with arithmetic operations, to create expressions with different numerical values. Such expressions are called **1-2-3-4 expressions.** For instance, $1 + (2 + 3) \cdot 4$ is a 1-2-3-4 expression for the number 21.

Unlike the original problem, this assignment involves negative as well as positive integers, so read the instructions carefully.

## *The Task*

Create as many 1-2-3-4 expressions as you can for each of the numbers from –20 to 20, using the rules outlined here.

*Continued on next page*

# *The Rules*

There is one essential rule for forming 1-2-3-4 expressions.

- You must use each of the digits 1, 2, 3, and 4 exactly once.

The digits can be combined using any of these methods.

- You may use any of the four basic arithmetic operations—addition, subtraction, multiplication, and division (according to the order-of-operations rules).

- You may use exponents.

- You may use radicals or factorials.

- You may juxtapose two or more digits to form a number such as 12.

- You may use parentheses and brackets to change the meaning of the expression.

- You may use a negative sign in front of any of the digits 1, 2, 3, or 4. For example, $-3 \cdot (4^2 - 1)$ is a 1-2-3-4 expression for the number $-45$. (This method was not included in the original problem.)

*Note:* You may *not* just put a negative sign in front of an entire expression. For example, $-(3 + 4! + 1 - 2)$ is *not* a legitimate 1-2-3-4 expression for $-26$, even though $-(3 + 4! + 1 - 2)$ is equal to $-26$ and uses each digit exactly once. You can only put the negative sign in front of an individual digit.

### Days 6–12

# Keeping Things Balanced

*This page in the student book introduces Days 6 through 12.*

What is an equation? What does it mean to solve an equation? *Homework 6: The Mystery Bags Game* introduces a simple game using a pan balance that will be used throughout the unit as a way to think about these questions.

Over the next few days, you'll review ideas about substitution and order of operations. You'll work with families of algebraic expressions and use functions to represent situations.

For now, trial and error will be one of the main tools for solving equations, but you'll also begin using graphs and seeing how valuable a graphing calculator can be for solving equations.

*Jeff Trubitte and Rafael Pozos created two different outcomes for "POW 2: Tying the Knots."*

Interactive Mathematics Program                                                                    19

# *Similarity Equations*

Students
review
negative
numbers and
use algebra to
represent
situations.

## Mathematical Topics

- Reviewing the arithmetic of integers
- Using algebra to represent situations involving similar triangles
- Solving equations using guess and check
- Developing a pan-balance model for solving problems

## *Outline of the Day*

### In Class

1. Discuss *Homework 5: 1-2-3-4 Puzzle with Negatives*
   - Encourage use of the hot-and-cold-cube model

2. Discuss *Lamppost Shadows* (from Day 5)
   - Focus on students' methods for solving equations and on the interpretation of solutions

3. Introduce *Homework 6: The Mystery Bags Game*
   - Have students read the introduction, describe the situation in their own words, and do one or two examples

### At Home

*Homework 6: The Mystery Bags Game*

## 1. Discussion of *Homework 5: 1-2-3-4 Puzzle with Negatives*

*"Who can summarize the model of the hot-and-cold cubes?"*

You can ask various heart card students to share some of their 1-2-3-4 expressions.

As needed, review the rules for working with negative numbers. If students seem confused, have them think back to the hot-and-cold-cube model from *Patterns* (see *The Chef's Hot and Cold Cubes* on Day 13 of that unit) and try to explain the numerical value of 1-2-3-4 expressions using that model. You may also want to have one or two students summarize how the model works and give some examples illustrating the ideas.

Keep in mind that some students may wish they could simply rely on rules. You can remind them that understanding "the big picture" provided by the model can help them if they forget the rules and will also help them avoid applying the rules incorrectly.

# 2. Discussion of *Lamppost Shadows*

The primary focus of the discussion of *Lamppost Shadows* should be on how students solved the equations and interpreted their solutions in terms of the problem.

As needed, continue the review of similarity that began with Question 5 of *Memories of Yesteryear* from Days 2 and 3. The next section, "Using similar triangles," provides some ideas to use for reviewing similar triangles. You may find it easier to develop the equations for Questions 1 and 2 before looking at the process of solving either equation.

During the discussion, you may want to bring up some of these points as well, although you should take care to maintain the central focus.

- Students will probably get only an approximate answer for Question 2 (using trial and error). You can use this fact to motivate the use of algebraic methods to obtain exact answers.

- You can use the discussion about the relationship between the equation and the problem to bring out that the degree of exactness needed in solving the *equation* depends on the degree of exactness desired in solving the *problem*.

- In Question 1a, students were asked to make an estimate from the diagram. You can look for an opportunity to bring out that such an estimate only makes sense if the diagram is roughly to scale.

## • *Using similar triangles*

*"Why are the two triangles similar?"*

In developing the equation for Question 1, be sure students recognize that these are similar triangles and that the proportionality of similar triangles is applied to get the equation $\frac{S}{S+30} = \frac{5}{25}$. If necessary, ask leading questions like, "Why are the two triangles similar?" rather than insisting that students develop the equation entirely on their own.

*As a hint: "What's the ratio of the horizontal sides of the two triangles?"*

It might be helpful to ask students how to express the ratio of the horizontal sides in the two triangles. Much as the phrase "ox expressions" was used in the Year 1 unit *The Overland Trail*, you might call the ratio $\frac{S}{S+30}$ a "shadow expression" or a "similar-triangle expression." Students can first focus on the individual ratios and then simply set them equal to each other to create an equation.

*"What must you change in the equation for Question 1 to make it fit Question 2?"*

You should also discuss briefly how to adapt the equation from Question 1 to the circumstances of Question 2. Make students aware that the geometric principles in the two problems are the same and that they merely need to change the numbers.

*Note:* Students typically will use the small "shadow triangle" and the large "lamppost triangle" as the similar triangles. But others may remember (or reinvent) the approach from *More Triangles for Shadows* on Day 21 of *Shadows,* creating a diagram like the one shown here and comparing the two smaller triangles to get *S* by itself in the equation.

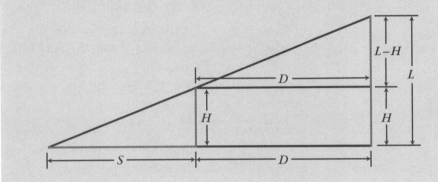

## • *Solving the equations*

After briefly discussing where the equations come from, focus on how to solve them. Because students may not have much experience with solving equations, you may need to emphasize the basic notion that "solving an equation" simply means finding a value that can be substituted for the variable to make the equation true.

There is no "right" or "wrong" way for students to actually find solutions. They will probably have used primarily guess-and-check to solve these equations, although some may know or have developed more sophisticated techniques.

You might list any strategies or techniques that students use. If the concept of cross-multiplying or using the distributive property comes up, tell students that they will be exploring the ideas behind those techniques later in this unit.

If students use methods other than guess-and-check, remind them to *check* their answers by seeing if they actually fit the equation or at least come close. (If they used guess-and-check, this reminder is unnecessary.)

*Note:* Now that students are in class rather than at home, they may want to use tables or other shortcuts on their graphing calculators to simplify the process of replacing $S$ with various values. This unit explores graphical techniques later on, so the use of graphs to solve the equations need not be discussed here unless the issue is raised by students.

The answer to Question 1c, exactly 7.5 feet, is one that students likely will find by guessing. However, students probably will not find the exact answer to Question 2c, $6\frac{6}{19}$ feet (approximately 6.32 feet), without algebraic manipulation.

### • *Solving the equation versus solving the problem*

Bring out the fact that solving an equation that comes from a problem is not necessarily the same as solving the original problem. An equation is an abstraction of the problem, and students should always put their solution of the equation back into the context of the problem. You may want to refer to this process as a "reality check."

Whatever methods students use for solving the equations, they should understand the significance of their solutions. For example, if you are discussing the equation from Question 2, $\frac{S}{S+20} = \frac{6}{25}$, and students come up with $S \approx 8$, ask them what this number means. They should be able to articulate that this means that the shadow is about 8 feet long.

*"Does this seem to be a reasonable answer to the problem?"*

Ask students if this seems to be a reasonable answer to the problem. They should get into the habit of thinking about problems intuitively before they dive into the manipulations. (This is a good place to relate Question 1a to other issues regarding approximation.)

*Comment:* Don't let students forget that they need to include *units* (feet, in these problems) with the numbers that represent measurements.

## 3. Introduction to Homework 6: The Mystery Bags Game

Tonight's assignment, *Homework 6: The Mystery Bags Game,* introduces a situation based on a pan-balance model. The game is a playful introduction to the principles of balancing equations, and the balance metaphor will be used later in the unit as a model for more formal thinking about solving equations.

The situation is fairly straightforward, but you should take a few minutes to introduce it, because it will play an important role in the unit. You can begin by having students read the activity through Question 1 and then having one or two students describe what's going on in their own words. The key features to establish are

- each mystery bag weighs the same amount
- the pans on the two sides balance exactly

Then have someone solve Question 1. (If students understand the situation, this should involve nothing more than simple division. If they suggest other solutions, they probably are confused about what's going on.)

As time allows, do one or two more examples, either directly from the assignment or similar ones of your own creation.

## Homework 6: The Mystery Bags Game

The last few questions of *Homework 6: The Mystery Bags Game* will push students to think more deeply about the mystery bags model.

***Travis Perez, Heather Sears, and Robert Rice discuss how their classwork "Back to the Lake" emphasizes the relationship between situations, rules, tables, and graphs.***

# Homework 6                                    The Mystery Bags Game

Do you remember the king in the "Bags of Gold" POWs? Well, he doesn't let the gold out of his sight anymore. But it can get very boring watching gold all day, so he has the court jester make up games for him to pass the time.

The game the king loves best is the mystery bags game. First, the jester takes one or more empty bags and fills each bag with the same amount of gold. These bags of equal weight are called the "mystery bags." Next, the jester digs into his collection of lead weights. He takes out his pan balance and places some combination of mystery bags and lead weights on the two pans so that the two sides balance.

The game is to figure out the weight of each mystery bag.

*Continued on next page*

# *Your Task*

The game may sound rather easy, but it can get very difficult for the king. See if *you* can win the mystery bags game in the various situations described here by figuring out how much gold there is in each mystery bag.

Explain how you know you are correct. You may want to draw diagrams to show what's going on. (The picture at the beginning of this assignment shows what the situation in Question 1 might look like.)

1. There are 3 mystery bags on one side of the balance and 51 ounces of lead weights on the other side.

2. There are 1 mystery bag and 42 ounces of weights on one side, and 100 ounces of weights on the other side.

3. There are 8 mystery bags and 10 ounces of weights on one side, and 90 ounces of weights on the other side.

4. There are 3 mystery bags and 29 ounces of weights on one side, and 4 mystery bags on the other side.

5. There are 11 mystery bags and 65 ounces of weights on one side, and 4 mystery bags and 100 ounces of weights on the other side.

6. There are 6 mystery bags and 13 ounces of weights on one side, and 6 mystery bags and 14 ounces of weights on the other side. (The jester could get in a lot of trouble for this one!)

7. There are 15 mystery bags and 7 ounces of weights on both sides. (At first, the king thought this one was easy, but then he found it to be incredibly hard.)

8. The king wants to be able to win easily all of the time, without calling you in. Therefore, your final task in this assignment is to describe in words a procedure by which the king can find out how much is in a mystery bag in any situation.

# DAY 7 *Equations for Bags*

## Mathematical Topics

- Developing a balance model for solving problems
- Representing the balance model algebraically

## Outline of the Day

### In Class

1. Discuss *Homework 6: The Mystery Bags Game*
   - Have students describe their general strategy in words
2. Introduce algebraic representation of the mystery bags game

- Have students write symbolic representations of their solutions to mystery bags problems

### At Home

*Homework 7: You're the Jester*

---

### Discuss With Your Colleagues

#### A Metaphor for Equations

**Yesterday, you introduced students to the mystery bags game. How did they react to this idea, or how have students reacted in the past? Do they resist the use of the metaphor as childish, or does the metaphor really work for them? Also, what are the advantages and disadvantages of different approaches to developing rules for finding equivalent equations?**

## 1. Discussion of *Homework 6: The Mystery Bags Game*

Have students share their explanations and answers in groups while you check off their homework. Then bring the class together and have diamond card members of different groups present Questions 1 through 7.

---

When you bring the whole class together, keep in mind that later in the unit, the work with equations will become more algebraic and abstract.

For now, the discussion needs to be as concrete as possible. Suggest that students use diagrams to help themselves visualize the situations.

## • *Questions 1 through 5*

*On Question 1:
"Exactly where
does the number 17
come from?"*

Questions 1 through 5 will probably be fairly straightforward, but it's worth going through each of them in detail in order to establish the intuitive principles for working with equations. For example, on Question 1, the presenter will probably say that because the three bags weigh 51 ounces, each bag must weigh 17 ounces. Have the presenter state explicitly where the number 17 came from to get her or him to articulate the process of division.

On Question 2, have the presenter state clearly that he or she used subtraction to determine a weight of 58 ounces for the single mystery bag. Try to get students to visualize this as removing 42 ounces of lead weights from each side.

Question 4 is the first example involving mystery bags on both sides. Here you want to get the presenter to talk explicitly about removing three bags from each side. This concrete visualization and manipulation of "the unknown quantity" can be helpful later on in keeping students from making mistakes such as simplifying $4x - 3x$ as 1 (as if the $x$'s cancel out).

## • *Questions 6 and 7*

Questions 6 and 7 present special situations. Students presumably will see intuitively that the pans can't balance in Question 6. That is, if the mystery bags all weigh the same amount, then this situation is impossible.

It may be a bit more difficult for students to articulate what's going on in Question 7. They should see that the pans will balance no matter what the weight of each mystery bag (as long as the bags all weigh the same). Bring out that the king really has no way to figure out the weight of a mystery bag in this situation.

## • *Question 8*

You may want to have several volunteers present their descriptions of a general procedure. They might come up with a general strategy like this:

- If there are mystery bags on both pans, remove the same number of mystery bags from each pan so there are mystery bags on only one pan.

- If there are weights on both pans, remove the same amount of weight from each pan so there are weights on only one pan.

- When there are only bags on one pan and only weights on the other, divide the amount of weight by the number of mystery bags.

You might ask what could go wrong with such a strategy in situations like those in Questions 6 and 7. Students should see that the difficulty comes from having the same number of bags on each side initially. You might get them to follow the process through to see that the final step would involve dividing by zero.

## 2. Representing the Mystery Bags Algebraically

***"How can you represent Question 1 algebraically?"***

Next, ask the class how to write an algebraic representation of the situation in Question 1, using $M$ for the weight of a mystery bag. They will probably come up with either $M + M + M = 51$ or $3M = 51$.

***"How can you represent the other problems algebraically?"***

Then have students work in groups to come up with algebraic representations of the rest of the problems. When the groups are done, assign one problem to each group. Have the club card member present the group's equation to the whole class.

***"How do the arithmetic operations compare with your general strategy?"***

Discuss with the class how the arithmetic operations compare with the steps in the general strategy developed earlier. Ask the class how they could write algebraically the steps they followed in last night's homework.

For instance, Question 5 might look like this:

$$11M + 65 = 4M + 100$$
$$\underline{-4M \qquad\qquad -4M}$$
$$7M + 65 = \qquad 100$$
$$\underline{\qquad -65 \qquad\qquad -65}$$
$$7M \quad = \qquad 35$$

$$M = 35/7$$
$$M = 5$$

*Note:* Some students may make the mistake of thinking that $11M - 4M$ is equal to 7, believing that they can "cancel out" the $M$s. If this is the case, you will need to address the problem. You can use the model of mystery bags and lead weights to clarify this. That is, students should see that removing 4 mystery bags from the original 11 will leave 7 mystery bags, not 7 ounces of lead weight.

### • *More algebra and mystery bags problems*

Have groups write steps like the ones just shown for each problem on last night's homework and if you have time, let students make up their own mystery bags problems and exchange them with one another.

## Homework 7: You're the Jester

Tonight's homework gives practice in solving linear equations using the pan-balance model.

# Homework 7  You're the Jester

1. Here are some simple equations that might have come from mystery bags games. Solve each equation for $M$, which represents the weight of each mystery bag.

   a   $M + 16 = 43$

   b   $12M = 60$

   c.  $27 + 9M = 90$

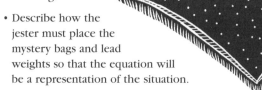

2. The equations in the next group are a bit more complicated. For each equation, do two things.

   • Describe how the jester must place the mystery bags and lead weights so that the equation will be a representation of the situation.

   • Find the weight of one mystery bag and explain how you got the answer.

   a.  $5M + 24 = 51 + 2M$

   b.  $43M + 37 = 56M + 24$

   c.  $12M + 15 = 5M + 62$

3. Make up two equations of your own like those in Question 2. Describe the jester's setup for each of your equations, and find the weight of one mystery bag in each case.

# Substitution and Order of Operations Revisited

*Students continue work with balancing and review substitution and order of operations.*

## Mathematical Topics

- Representing a balance model algebraically
- Substitution and evaluation with algebraic expressions
- Reviewing order of operations

- - - - - - - - - - - - - - - - - - - - - - - - - - - - - - - - - - -

# Outline of the Day

## In Class

1. Discuss *Homework 7: You're the Jester*

   - Do some examples from Questions 1–3
   - Do a more complex example
   - Develop the general solution

2. Review the process of substitution and the conventions for order of operations

3. Refer students to *Substitution and Evaluation*

   - This reference material summarizes the two-step process for evaluating algebraic expressions

## At Home

*Homework 8: Letters, Numbers, and a Story*

## 1. Discussion of *Homework 7: You're the Jester*

You might put a problem like the following on an overhead transparency for students to work on while you check off who completed the homework.

$$5M + 27 + 3M = 23 + 6M + 54$$

You may want to give students more than one problem to do. In any case, include at least one at this level of complexity—that is, just a bit harder than those on the homework.

Begin the actual discussion by going over some of the problems from Questions 1 and 2. Use your judgment about how many of these problems to discuss. If students seem to need more practice, you may want to let some students share the problems they made up for Question 3.

Then review the more complex in-class problem. Some students may combine terms to express the problem as $8M + 27 = 6M + 77$. The idea of combining $5M$ and $3M$ as $8M$ may seem simple in the context of the mystery bags problem, but students may have difficulty with this process in a purely algebraic setting. Use the mystery bag context to go over the process, bringing out that "five of something" plus "three of the same thing" is always "eight of those things." (The formal phrase "combining like terms" to describe this step is introduced on Day 14.)

Other students may go straight to the stage of "removing bags" and so on, without first combining terms. For example they may simplify the original equation to $5M + 27 = 23 + 3M + 54$ (by removing three bags from each side) and continue with similar steps to get to the final answer.

Either approach is fine, and you should encourage students to explore different ways to look at the process.

## 2. Reviewing Substitution

*"How do you check if you have the right answer for a mystery bags equation?"*

Ask students what the best way is to be sure that they have the right answer for a mystery bags equation. They will probably say something about replacing $M$ with the number that they think is the correct value.

Have a student illustrate how this works using one of the examples from Question 2. For instance, in Question 2a, where the mystery bag weighed 9 ounces, a student might say something like, "If $M$ is 9, then $5M + 24$ is $45 + 24$, which is 69, and $51 + 2M$ is $51 + 18$, which is also 69."

Point out that students can think of this process in two stages:

- **Substitution,** which is simply replacing a variable with a number

- **Evaluation,** which is getting a single number out of the substitution process

Tell students that either term is sometimes used for the entire process but that it is sometimes helpful to use separate terms for the two steps, so that the "algebra" part is distinct from the purely "arithmetic" part.

*Note:* The concept of substitution, including this two-step process, was introduced in the Year 1 unit *The Overland Trail,* as were the rules for order of operations. Both of these ideas, however, are of sufficient importance that they should be reviewed here.

### • *Reviewing order-of-operations rules*

A key part of the substitution/evaluation process is proper application of the rules for order of operations. To begin a review of the order-of-operations rules, you might look at the expression $51 + 2M$ (from Question 2a) and ask students why that doesn't come out to 477 when $M$ is 9, arguing that $51 + 2$ is 53 and $53 \cdot 9$ is 477.

*"What are the rules for order of operations?"*

Students should recall that certain conventions dictate which operation to do first—in this case, multiplication before addition. Ask students to reconstruct the order-of-operations conventions, and post a chart showing the order of priority for simplifying arithmetic expressions:

1. Parentheses

2. Exponents

3. Multiplication and division (equal priority) from left to right

4. Addition and subtraction (equal priority) from left to right

After clarifying these rules, you might return to the example of substituting $M$ into the expression $51 + 2M$ and focus on the multiplication sign implied between 2 and $M$. You can suggest to students that they develop the habit of putting parentheses around variables to indicate multiplication and then keeping the parentheses around the number being substituted. For example, they might write $51 + 2M$ as $51 + 2(M)$ before substituting and then write the substitution step as $51 + 2(9)$.

### • *Some more examples*

You may want to have students do a few more examples of substitution and evaluation, both to clarify any issues related to the order-of-operations rules and to provide a context for continued work with negative numbers. You can use tonight's homework assignment as a model for the type of problem to propose.

## 3. For Reference: *Substitution and Evaluation*

Refer students to the description of the two-step substitution process in *Substitution and Evaluation.* (This is reference material for students. You do not need to go over it explicitly, because it merely summarizes ideas already discussed.)

# Homework 8:
## Letters, Numbers, and a Story

Part I of tonight's homework is a routine exercise on substitution, with expressions that are likely to reveal common mistakes. The assignment emphasizes the two-stage substitu-tion/evaluation process. Part II continues students' work with the relationship between equations and situations, as in *Homework 3: You're the Storyteller.*

You may wish to point out to students in advance that Part I of this assignment illustrates some of the many different phrases that are used to express the idea of substitution, and emphasize that these phrases all mean the same thing.

# Substitution and Evaluation

You often need to find out what the numerical value of a particular algebraic expression would be if you replaced the variable with a number. This happens a lot in the guess-and-check approach to solving equations.

It's useful to identify and name two separate parts of this process of getting numerical values from algebraic expressions.

- **Substitution** is the step of replacing the variable with a number.

- **Evaluation** is the step of getting a single number from the result of the substitution step.

For example, consider the expression $x^2 + 5x - 3$. Suppose you wanted to see what would happen if $x$ were equal to 7.

*Continued on next page*

Interactive Mathematics Program                                              23

**Substitution:** In the substitution step, you simply replace each occurrence of the variable with the number 7, as shown here. Recall that 5(7) means 5 *times* 7.

Notice that the number 7 has been placed within parentheses in each case for clarity. This isn't always necessary, but it helps prevent mistakes, such as getting 57 instead of $5 \cdot 7$.

**Evaluation:** The evaluation step turns the numerical expression $(7)^2 + 5(7) - 3$ into a single numerical value. As shown here, you might first replace $(7)^2$ with 49 and 5(7) with 35, then add $49 + 35$ to get 84, and finally subtract 3 from 84 to get the final result.

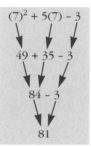

Here are some useful ways to express this overall process in words.

"The value of the expression $x^2 + 5x - 3$ for $x = 7$ is 81."

"Substituting 7 for $x$ in the expression $x^2 + 5x - 3$ gives the value 81."

"Evaluating $x^2 + 5x - 3$ at $x = 7$ gives 81."

Keep in mind that in doing the evaluation step, you need to follow the rules for order of operations. By convention, we simplify expressions according to this sequence:

1. Parentheses

2. Exponents

3. Multiplication and division (equal priority) from left to right

4. Addition and subtraction (equal priority) from left to right

## Warning: The Missing Multiplication Sign

According to the rules for order of operations, we apply an exponent before we multiply. For example, the numerical expression $3 \cdot 7^2$ means $3 \cdot 49$, and not $21^2$.

This rule also governs algebraic expressions, but many errors in the substitution/evaluation process occur because we leave out multiplication signs in algebraic expressions. For example, in the expression $3x^2$, there is a "missing multiplication sign" between 3 and $x^2$. In other words, $3x^2$ is shorthand for $3 \cdot x^2$.

Therefore, $3x^2$ means $3 \cdot (x^2)$ and not $(3 \cdot x)^2$. You may find it helpful to insert parentheses or explicit multiplication signs into algebraic expressions in order to be clear about what they mean.

# Homework 8

# Letters, Numbers, and a Story

## *Part I: Substitution and Evaluation*

Evaluate each of the eight expressions shown here according to these two steps.

- Replace the variable with the value shown, writing the resulting expression in complete detail.

- Compute the numerical value of the expression you get in the first step.

Be sure to insert parentheses or multiplication signs where needed.

*Note:* The instructions in Questions 1 through 8 illustrate some of the many ways by which the process of substitution is described. In each case, you should use both steps.

1. Evaluate $5 + 6q$ at $q = 9$.

2. Find the value of $3z + 20$ when $z = -8$.

3. Get the numerical value of $15 - 4x$ for $x = -1$.

4. Evaluate $3t^2 + 7$ if $t = -2$.

5. What is $-r^2$ when $r = 8$?

6. Find $-z^2$ with $z = -6$.

7. Substitute $k = 3$ into $3 \cdot 2^k + 5$.

8. Evaluate $3x^3 + (4x)^2$ using $x = 5$.

## *Part II: Make Up a Story*

Create a situation and then write a question about your situation so that solving the equation $4(x + 3) = 40$ will answer your question. Be sure to identify what the variable $x$ represents in your situation. Once you solve the equation, explain what the solution means in terms of your situation.

# DAY 9 Catching Up

*Students use numerical examples to develop an equation for a situation involving rates.*

## Mathematical Topics

- Continuing a review of order of operations and substitution
- Developing an equation by considering numerical examples

# Outline of the Day

## In Class

1. Discuss *Homework 8: Letters, Numbers, and a Story*
2. *Catching Up*
   - Students use a variable to represent an unknown and develop an equation by considering numerical examples

3. Discuss *Catching Up*
   - Focus on modeling the equation after the numerical examples

## At Home

*Homework 9: More Letters, Numbers, and Mystery Bags*

## Discuss With Your Colleagues

### Equations from Examples

Today's activity, *Catching Up,* uses a detailed sequence of steps to help students develop an equation based on the numerical patterns involved in testing a guess. Does this approach seem to work? Do some students find it too slow? How can you help students develop their own best methods for writing equations for word problems?

# 1. Discussion of *Homework 8: Letters, Numbers, and a Story*

You can assign one of the problems from Part I to each group and have one group member put each solution on the board. You can put up all of the problems at once (if board space allows). Remind students to show all steps of the substitution/evaluation process.

You can also have each group be responsible for checking that another group's problem has been done correctly. You need not go over these problems unless there are questions about or challenges to the work that has been displayed.

*"How can you explain the problem using the hot-and-cold-cube model?"*

If students are having difficulty with or are confused about issues of sign, have them go back to the model of hot and cold cubes rather than get into disputes about who remembers the rules correctly. (You may want to have them take out their portfolios from the Year 1 unit *Patterns* and review their work on *Homework 14: You're the Chef*.)

For example, in Question 4, students have to evaluate $3(-2)^2 + 7$. If they aren't sure whether $(-2)^2$ should be positive or negative, you can refer to the hot-and-cold-cube model to help illuminate why the product of two negative numbers is positive.

> For instance, they might say, "The product $(-2) \cdot (-2)$ is like the chef taking out two bunches with two cold cubes in each bunch. The temperature would go up 4 degrees, so the product is 4."
>
> *Note:* A supplemental problem, *Preserve the Distributive Property*, uses an algebraic approach to explain this rule of signs. You might want to assign that activity to students after about Day 16.

While students are putting solutions up on the board and checking each other's work on Part I, you can circulate and look for students who seem to have come up with good situations for the equation in Part II. Once the work on substitution is completed, have these students share their work on Part II.

# 2. Catching Up

> This activity will give students more experience using a variable to represent an unknown. It is designed to encourage them to use numerical examples to develop equations.

# 3. Discussion of *Catching Up*

You might have various spade card students each discuss a different part of the activity. Be sure to encourage alternate explanations as well.

For Questions 1a, 1b, 2a, and 2b, students may simply calculate miles traveled as this product:

(# of miles per hour) • (# of hours per day) • (# of days)

Others may first compute the number of miles per day separately and then multiply by the number of days. Try to bring out any variations in how students handle these specific cases.

Questions 1c and 2c are important in building toward the use of an equation. Be sure that students explain how they know that the Sawyers have not yet caught up.

For example, in Questions 1a and 1b, students will have found that after 6 days, the Sawyers have traveled 180 miles while the wagon train has traveled 144 miles. They might give as their explanation in Question 1c that the Sawyers have gained only 36 miles. You can ask how they got that value and how that tells them that the Sawyers haven't caught up.

Other students may have reasoned that after 6 days, the wagon train is 144 + 120 = 264 miles from where the Sawyers originally were while the Sawyers are only 180 miles from that point.

Be sure that the specific examples from Questions 1 and 2 are clear before moving on to a discussion of Question 3.

Students should get $3 \cdot 10 \cdot N$ (or something equivalent, such as $30N$) for Question 3a and $3 \cdot 8 \cdot N$ (or an equivalent) for Question 3b. Their approach on Questions 1c and 2c will probably determine how they express the equation. For example, if they focused on the distance the Sawyers gain, they might write the equation as $30N - 24N = 120$. If they focused on having the total distance that the Sawyers and the wagon train travel be the same, they would be more likely to write $24N + 120 = 30N$.

### • *Solving the equation and interpreting the result*

Try to find students with different forms of the equation for Question 3c and use these as opportunities to review ideas based on the mystery bags metaphor. For instance, if a student wrote the equation in the form $30N - 24N = 120$, you can review the idea of combining like terms to rewrite this as $6N = 120$. (See the discussion of *Homework 7: You're the Jester* for ideas on how to use the mystery bags metaphor to do this.) If a student wrote the equation as $24N + 120 = 30N$, you can review the mystery bags approach to solving equations by "removing bags from both sides."

*"How do you
know whether
N = 20 solves the
problem?"*

You can also use this as an opportunity to apply substitution by asking
students how they can be sure that $N = 20$ is the solution to the equation. Be
sure that they not only verify that $N = 20$ fits the equation but also go back
to the problem to confirm that it works. Essentially, this means doing
something similar to Questions 1 and 2 but using the value of 20 days.

*Comment:* Some students may point
out that the Sawyers gain 6 miles
each day, because they travel an
extra 2 hours at 3 miles per hour.
They may see that they can simply
divide the distance to be made up
(120 miles) by this daily rate of gain
to get the result of 20 days. If
students do in fact suggest this
approach, you should acknowledge
the insight it reveals while at the
same time bringing out the value of
the more deliberate guess-and-check
approach suggested by Questions 1
and 2.

# *Homework 9: More Letters, Numbers, and Mystery Bags*

This assignment continues the work on substitution and on the mystery bags
metaphor.

# Catching Up

The Sawyer family is 120 miles behind the rest of the wagon train and needs to catch up. Both the wagon train and the Sawyers' wagon travel at about 3 miles per hour.

The Sawyer family realizes that in order to make up the difference, they will have to travel more hours each day. They know that the wagon train travels 8 hours each day. Therefore, the Sawyer family decides that they will travel 10 hours every day.

1. a. How far will the Sawyer family travel in 6 days?

   b. How far will the wagon train travel during those 6 days?

   c. Will the Sawyer family catch up in 6 days? Explain your answer.

2. a. How far will the Sawyer family travel in 15 days?

   b. How far will the wagon train travel during those 15 days?

   c. Will the Sawyer family catch up in 15 days?

3. Use $N$ to represent a number of days.

   a. Write an expression for how far the Sawyer family will travel in $N$ days.

   b. Write an expression for how far the wagon train will travel during those $N$ days.

   c. Write an equation that states that the Sawyer family has caught up to the wagon train.

   d. Solve your equation from Question 3c and interpret the result.

# Homework 9   More Letters, Numbers, and Mystery Bags

## *Part I: Substitution and Evaluation*

1. As in *Homework 8: Letters, Numbers, and a Story*, evaluate each of these expressions showing these two steps.

   • Replace the variable by the value shown.

   • Compute the numerical value of the resulting expression.

   a. Find the value of $5a^2 + 3a + 4$ for $a = -2$.

   b. Evaluate $-r^3 + 2r^2 + 4r$ when $r = -3$.

   c. What is $(m^2 + 2)(m - 1) - (m - 1)(m^2 + 3)$ if $m = -7$?

   d. Substitute $c = -5$ into the expression $6(c + 4) - 3c(c - 1)$.

   e. Get the numerical value of $(v + 5)(v^2 - 4) - (v - 5)(v^2 + 4)$ at $v = 7$.

   f. Evaluate $y^3 + (2y)^2$ at $y = -5$.

   g. Find $2r^2 - 5r + 9$ with $r = -6$.

*Continued on next page*

2. Make up two substitution examples of your own using these steps.

- Decide what letter to use as the variable.

- Make up an expression using that variable.

- Pick a number to substitute as a value for the variable.

- Substitute the number for the variable and then evaluate the resulting expression.

In one of your examples, substitute a positive number. In the other example, substitute a negative number.

## *Part II: More Mystery Bags*

3. Solve each of these equations and explain your work using the mystery-bags model.

a. $15M + 43 = 37M + 12$

b. $52x + 19 = 23 + 16x$

c. $5t + 12t + 13 = 8t + 19$

d. $9a + 6 + 3a + 7 = 10a + 21 + 6a$

e. $3r + 4 + 2r = 7 + r + 4r$

# DAY 10 *Back to the Lake*

*Students use an algebraic expression to study a family of related equations.*

## Mathematical Topics

- Continuing work with substitution
- Continuing work with solving equations using a balance metaphor
- Creating functions for situations by considering numerical examples
- Relating functions and equations
- Reviewing function notation

## Outline of the Day

### In Class

1. Select presenters for tomorrow's discussion of *POW 1: A Digital Proof*

2. Discuss *Homework 9: More Letters, Numbers, and Mystery Bags*

3. *Back to the Lake*
   - Students use an algebraic expression to study a family of related equations

4. Discuss *Back to the Lake*
   - Emphasize the different representations—table, equation, and graph

5. Review function notation using *Back to the Lake*

### At Home

*Homework 10: What Will It Answer?*

## Discuss With Your Colleagues

### Can You Give Them the Formula?

**In Question 4 of tonight's *Homework 10: What Will It Answer?* students are given the physics formula $F = ma$ without any explanation. Aren't IMP students supposed to figure out formulas for themselves? Is it appropriate to provide this formula (and others, as in *Homework 26: More Variable Solutions*) without discussing where the formula comes from and what it means?**

## 1. POW Presentation Preparation

Presentations of *POW 1: A Digital Proof* are scheduled for tomorrow. Choose three students to make POW presentations and give them overhead transparencies and pens to take home for preparing presentations.

## 2. Discussion of *Homework 9: More Letters, Numbers, and Mystery Bags*

You can handle the discussion of Question 1 of *Homework 9: More Letters, Numbers, and Mystery Bags* similarly to the discussion of Part I of *Homework 8: Letters, Numbers, and a Story*. If students seem to need more work on substitution, you can have several students share their examples from Question 2.

Use your judgment about how much time to spend on Questions 3a through 3d. Students should become fairly comfortable with solving these simple linear equations so that they can move on to more complex examples later in the unit.

You should definitely discuss Question 3e, however. Students will probably recall a similar situation from *Homework 6: The Mystery Bags Game*. Be sure they realize that this is an equation with no solution and that "no solution" is sometimes the best response.

## 3. *Back to the Lake*

Students have been writing equations and expressions, and they have substituted and evaluated. In *Back to the Lake,* students use several variations on a specific situation to develop a function describing the general situation. They then use an In-Out table and a graph to gain further insight into the situation.

One goal of the unit is for students to see situations, rules, tables, and graphs as different ways to represent the same thing. This is done by providing opportunities for them to see these forms together in different contexts. The interconnection among these different representations is a theme that was introduced in *The Overland Trail* and that will be developed throughout the four years of the IMP curriculum.

No specific introductory work is needed for this activity.

## 4. Discussion of *Back to the Lake*

You can begin by having different heart card students present Question 1 and some examples they used in creating their table in Question 2. Then turn to one or two presentations of Question 3. Students should come up with something like $d = \frac{6}{N}$.

Students might explain this in terms of the problem by pointing out that in every case, Yolanda does 6 miles of "around the lake" jogging.

### • *Question 4*

In connection with Question 4, you may need to review the convention that the *In* is represented along the horizontal axis and the *Out* is represented along the vertical axis. You can also review the terms *independent variable and dependent variable*. (These ideas and terms were introduced on Day 11 of *The Overland Trail* in Year 1.)

You may also want to discuss how to choose scales for the axes. Review the fact that the vertical and horizontal scales do not necessarily have to be the same, especially when they represent different kinds of quantities (here, number of laps and distance in miles).

Some students may have plotted individual points for Question 4, while others may have connected the points. You can bring out that Yolanda is actually only interested in points whose horizontal coordinate is a whole number.

*"What did you notice about the individual points you plotted?"*

In any case, however, ask what students notice about the individual points they plotted. One important observation is that the points do *not* lie on a straight line. In *Homework 11: Line It Up,* students will be examining the question of what types of expressions give linear graphs, so you might want to point out here that the function is **nonlinear.**

Students may also observe that the graph moves down as it goes to the right. You might ask students to explain this in terms of the situation. For example, they might say that the more times Yolanda goes around the lake, the smaller the size of the lake she needs.

*"What parts of the graph make sense in this problem?"*

You can also ask students what parts of the graph make sense in this problem. For example, they might point out that if $N$ is very big, the value of $d$ gets so small that it couldn't represent the distance around a lake.

### • Three representations

Bring out that students have created three different representations of the situation:

- An In-Out table (in Question 2)

- An equation or rule (in Question 3)

- A graph (in Question 4)

*"How are these representations connected?"*

Ask students to comment on the connections among these relationships. They should see that a row of numbers in the table represents the coordinates of a point on the graph and that these numbers will also fit the equation. You might also ask students which of these representations they find the most useful or the clearest. There is no "right answer" here—the goal is to provide another stimulus for students to think about these ideas.

## 5. Review of Function Notation

*"What 'big idea' are tables, rules, and graphs representing?"*

Ask students what general concept is being represented when they use any of these three approaches. As needed, remind students that In-Out tables, rules, and graphs are simply different ways to think about *functions*.

To review function notation, tell students that you're going to use the letter $f$ to represent the function relating $N$ to $d$. Then ask them how they could use this abbreviation in an equation. Several categories of response are possible, including

- a generic equation such as $d = f(N)$

- an equation based on the formula from Question 3, such as $f(N) = \frac{6}{N}$

- equations based on particular examples, such as $f(2) = 3$

You should elicit all three types of response from the class.

To get the first, you might ask for an equation using $d$, $N$, and $f$. To get the second, you might ask for a formula showing $f(N)$ in terms of $N$. To get the third, you might ask what $f(2)$ means and how one gets it.

Once you have examples of all three kinds of statements, ask questions to emphasize the connection between the formal notation and the situation. One type of question presents students with a statement in function notation and asks them to come up with a statement about the problem situation. For instance, ask them what $f(10)$ is and follow up by asking what the numerical result represents in terms of the situation.

Also give students a question about the situation and have them express it in function notation. For example, ask them how many times Yolanda would go around a 1.2-mile lake and then ask them to write an equation using $f$ that goes with this question. They might express this with the equation $f(N) = 1.2$.

Also have them write this equation without function notation. They should see that this is the same as writing $\frac{6}{N} = 1.2$.

*Comment:* Although these questions have simple answers, they do involve a level of comfort with the formalism of functions and equations that takes some time and practice for many students to develop.

## Homework 10: What Will It Answer?

In today's activity, *Back to the Lake,* students were asked to develop a function that would allow Yolanda to find the necessary distance $d$ around the lake once she decided on the number of times $N$ that she would jog around the lake. In other words, they wrote $d$ as a function of $N$, so that they could find $d$ if they knew $N$.

Thus, the equation for the function allows students to find the dependent variable if they know the value of the independent variable. In this assignment, they should see that such an equation can also be used in the other direction.

# *Back to the Lake*

The following problem is from *Homework 4: Running on the Overland Trail.*

> Yolanda jogged 2 miles to a lake, jogged twice around the lake, and then jogged 2 more miles home. Altogether she traveled 10 miles. How far is it around the lake?

In that particular situation, the distance around the lake was 3 miles.

Well, Yolanda always does a 10-mile jog, and she likes to go 2 miles to the lake and 2 miles back, but she gets tired of always going around the lake twice. Fortunately, Yolanda lives in Minnesota, and several lakes of varying sizes are 2 miles from her home. She would like to be able to choose a lake of the right size depending on how many times she wants to go around.

1. Suppose Yolanda wants to jog 2 miles to a lake, go *four* times around it, jog 2 miles home, and have that be a total of 10 miles. How big a lake should she choose? That is, how far should it be around the lake?

*Continued on next page*

Interactive Mathematics Program                                                    29

2. Now set up an In-Out table to describe situations like Yolanda's original jog and Question 1. In each case, Yolanda jogs 2 miles to a lake, goes some number of times around it, and jogs 2 miles home, for a total of 10 miles. The *In* for your table should be the number of times Yolanda goes around the lake, and the *Out* should be the distance around the lake.

   a. Use Yolanda's original situation for one row of the table and Question 1 for another. (For instance, for the original situation, the *In* would be 2, because she went around twice, and the *Out* would be 3, because the distance around the lake was 3 miles.)

   b. Create two more rows by choosing two other values for the *In*.

3. Now find a rule for your In-Out table from Question 2. Use *N* for the *In* and *d* for the *Out*, and write *d* as a function of *N*. In other words, write an equation in the form

$$d = \text{an expression involving } N$$

   so that Yolanda could substitute any value she wanted for *N* and find the size of the lake she needs.

4. Make a graph based on the In-Out table from Question 2, using your equation from Question 3 to find more points for the graph.

# Homework 10          What Will It Answer?

An important part of understanding what an equation or function means is knowing what types of questions it can be used to answer. For instance, the equation $A = s^2$ gives the area of a square ($A$) as a function of the length of its side ($s$).

The simplest use of this equation is to answer the question "What is the area of the square?" when you know the length of its side. For example, if you know that the length of a side is 5 inches, then you can substitute 5 for $s$ to find that the area of the square is $5^2$ (or 25) square inches.

The equation $A = s^2$ is even more powerful when you realize that it can also be used to answer a different type of question. That is, it can answer the question "What is the length of the side of the square?" if you know the square's area. For example, if you know that the area is 49 square inches, then the equation tells you that $49 = s^2$, which means that the side length must be 7 inches. (Why doesn't a solution of $s = -7$ make sense, even though it fits the equation $49 = s^2$?)

*Continued on next page*

1. The *perimeter* of a square is also a function of the length of a side.

    a. Choose variables and use them to write an equation describing this function. Be sure to state what your variables stand for.

    b. Give specific examples of two different types of questions that your equation can be used to answer. Also give the answers to your questions.

2. A movie theater charges $7 per ticket, and the theater's expenses are $500.

    a. Define appropriate variables and write an equation that gives the theater's profit as a function of the number of tickets sold. (Ignore such factors as the sale of refreshments.)

    b. Give specific examples of two different types of questions that your equation can be used to answer. Also give the answers to your questions.

Many principles in physics can be described in terms of functions. Questions 3 and 4 give two examples of this.

3. If an object is dropped and falls toward the ground, the distance it travels in $t$ seconds is given approximately by the equation $d = 16t^2$, where $d$ is the distance traveled, measured in feet. Come up with two different types of questions that this equation can be used to answer, and give the answers to your questions.

4. Newton's Second Law of Motion states that the force acting on an object ($F$) is equal to the object's mass ($m$) times the object's acceleration ($a$). In other words, the equation $F = ma$ gives the force as a function of mass and acceleration. What types of questions can this function be used to answer?

# POW 1 Presentations

*Students examine how equations can be used and present POW 1.*

## Mathematical Topics

- Seeing that a formula can be used to answer different types of questions
- Proving that a certain puzzle has only one solution
- Using a physical model to begin analysis of a combinatorial probability problem

## Outline of the Day

### In Class

1. Form new random groups
2. Discuss *Homework 10: What Will It Answer?*
3. Presentations of *POW 1: A Digital Proof*
4. Introduce *POW 2: Tying the Knots*
   - Students should see that there are three possible outcomes to the ritual

### At Home

*Homework 11: Line It Up*

*POW 2: Tying the Knots* (due Day 16)

### Special Materials Needed

- Sets of six pieces of string for a simulation of *POW 2: Tying the Knots.*

## 1. Forming New Groups

This is an excellent time to place students in new random groups. Follow the procedure described in the IMP *Teaching Handbook* and record the groups and suits for each student.

## 2. Discussion of *Homework 10: What Will It Answer?*

You can assign a problem or two to each group for presentation as you check off homework. You may want to overlap these (group 1 does 1 and 2, group 2 does 2 and 3, and so on) so that more than one group presents each problem.

*Comment:* The situations from Questions 2 and 3 of this assignment will be used in tomorrow's activity, *The Graph Solves the Equation,* so you may want to urge students to take notes as these problems are presented. You also may want to post the equations for these problems.

For Question 1a, the presenter should define variables (for example, $s$ = length of the side, $p$ = perimeter) and then give an equation for $p$ in terms of $s$ (presumably, $p = 4s$). For Question 1b, the presenter should give specific examples of two types of questions:

- If one knows $s$, one can find $p$.

- If one knows $p$, one can find $s$.

Be sure that specific examples are given. For example, students should say something like this:

- If you know that the square has sides of length 7, then you can substitute and get $p = 4 \cdot 7 = 28$.

- If you know that the perimeter is 60, then you can use the equation $60 = 4s$ to get $s = 15$.

*Note:* Without making a big deal out of it, you might point out that the algebra itself doesn't talk about units although the problem situation requires them. You might also want to look at the situation described in the introduction to the homework, where $s$ is in inches and $A$ is in square inches.

The situation for Question 2 may be slightly more difficult, but most students should come up with an equation like $p = 7t - 500$ (where $t$ = number of tickets sold and $p$ = profit). Again, the presenter should give specific examples in each direction.

In Question 3, bring out that students don't need to know the reason for the formula $d = 16t^2$, but you can mention that they will study the basis for this formula in the Year 4 unit *High Dive*. They should see that they can find either the distance an object falls (if they are given an amount of time) or the time involved (if they are given the distance the object falls).

Question 4 is more complex because there are three variables instead of two. The goal here is to get a set of statements like these:

- If you know the mass and the acceleration, then you can figure out the force.

- If you know the mass and the force, then you can figure out the acceleration.

- If you know the force and the acceleration, then you can figure out the mass.

## 3. Presentations of POW 1: A Digital Proof

Have the three selected students make their presentations. The presentations and discussion should focus on the proof that the solution given on Day 4 for *Is It a Digit?* is the only solution to the problem.

For your convenience, here is that solution again.

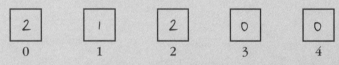

In one sense, the proof that the solution is unique consists of eliminating all other possibilities. However, because there are so many possible ways to fill in the boxes ($5^5$, or 3125), the other cases need to be eliminated in an organized and systematic way.

One idea that students might draw on both to find the solution and to prove that it is unique is that the numbers in the boxes must add up to 5. (You should justify this principle for yourself.) This condition limits the options considerably; in particular, no more than one "large" number can be used. Students can solve the problem in other ways, so you need not bring this up unless students do so.

Looking at simpler cases, in which there are only three or four boxes and the numbers 0 to 2 or 0 to 3, may also help illuminate why this solution is unique. It is interesting to see that the case of using boxes from 0 to 2 gives *no solutions* while the case of boxes from 0 to 3 gives *two solutions*.

This is an excellent problem for talking about proof. Students should recognize that saying "I couldn't find another solution" is a far cry from giving a proof that there are no others.

*Note:* A supplemental problem, *Ten Missing Digits,* poses a more complex version of the puzzle in *Is It a Digit?* and *POW 1: A Digital Proof.*

# 4. Introduction to POW 2: Tying the Knots

Let students read the POW through the description of the three stages. Then distribute six pieces of string to each pair of students and have the pairs work through the three stages (just once per pair). When the pairs are finished tying their knots, bring the class together and have each pair describe what it got.

The goal here is for students to see that there are three possible outcomes:

* Three small loops

* One small loop and one medium loop

* One large loop

In fact, it's fairly likely that none of the groups will get three small loops. If this outcome (or one of the others) doesn't occur, ask students if they think there are any other possible outcomes. It's probably best for the development of the problem if students leave class at least aware of the three possibilities. And you should clarify that a given number of loops represents the same outcome whether or not the loops are interlocked.

Students should realize that essentially they have just done Question 1 of the POW and that their task now is to find out the probabilities associated with each possible outcome. They can visualize this question as wanting to know approximately how many of each combination of loops to expect in a large class of a given size. Make sure they realize that an answer based solely on a simulation will not suffice, although they might want to use a simulation to test their analysis.

This POW is scheduled to be discussed on Day 16.

# Homework 11: Line It Up

Tonight's homework assignment continues the work with function notation and the connection between rules, graphs, and tables. It also will allow students to strengthen their intuitive understanding of what makes a function linear, in preparation for subsequent work in the unit with linear expressions, linear equations, and linear functions.

# POW 2                              *Tying the Knots*

Keekerik is an imaginary land where the people have an interesting three-stage ritual for couples who want to get married. Wandalina and Gerik are in that situation, so they go to the home of Queen Katalana to perform this ritual. Permission for them to marry as soon as they wish depends on the outcome of the ritual.

## Stage 1: Loose Ends Top and Bottom

The queen greets them and reaches into a colorful box to pull out six identical strings for the ritual. The queen hands the strings to Wandalina, who holds them firmly in her fist. One end of each string is sticking out above Wandalina's fist, and the other end of each string is sticking out below her fist.

## Stage 2: The Tops Are Tied

The queen steps to the side, and Gerik is called forward. He ties two of the ends together above Wandalina's fist. Then he ties two other ends above her fist together. Finally, he ties the last two ends above her fist together. The six ends below Wandalina's fist are still hanging untied.

## Stage 3: The Bottoms Are Tied

Now Queen Katalana comes forward again. Although she was watching Gerik, she has no idea which string end below Wandalina's fist belongs to which end above. The queen does the final step. She randomly picks two of the ends below and ties them together, then two more, and finally the last two. So Wandalina now has six strings in her fist, with three knots above and three knots below.

*Continued on next page*

Interactive Mathematics Program                                              33

## *Will They Be Able to Marry?*

Whether Wandalina and Gerik will be allowed to marry right away depends on what happens when Wandalina opens her fist. If the six strings form one large loop, then they will. Otherwise, they will be required to wait and repeat the ritual in six months.

With this in mind, think about these questions.

1. When Wandalina opens her fist and looks at the strings, what combinations of different size loops might there be?

2. What is the probability that the strings will form one big loop? In other words, what are the chances that Wandalina and Gerik will be able to marry right away?

3. What is the probability for each of the other possible combinations?

Although you may want to do some experiments to get some ideas about these questions, your answers for Questions 2 and 3 should involve discussion of the theoretical probability for each result, and not just experimental evidence.

## *Write-up*

1. *Problem Statement*

2. *Process:* Explain how you worked on this problem, including what experiments you performed and how you kept track of your results.

3. *Solution:* Give the probability for each possible outcome and explain how you determined each probability.

4. *Evaluation*

5. *Self-assessment*

# Homework 11                                  Line It Up

Probably the most important single type of function is the **linear function,** which can be defined as a function whose graph is a straight line.

1. Consider the function $f$ defined by the equation

$$f(x) = 3x + 2$$

   a. Find each of the values $f(1)$, $f(2)$, and $f(3)$.

   b. Use the results of Question 1a to complete this partial In-Out table for the function $f$.

| $x$ | $f(x)$ |
|-----|--------|
| 1   | ?      |
| 2   | ?      |
| 3   | ?      |

   c. Graph the points from your In-Out table.

   d. Do you think the graph of $f$ is a straight line? In other words, is $f$ a linear function? Explain your answer.

2. a. Plot the two points $(2, 3)$ and $(4, 7)$.

   b. Draw a straight line through your two points and find a third point on that line.

   c. Make an In-Out table like the one shown here, using your point from Question 2b as the third row

| In | Out |
|----|-----|
| 2  | 3   |
| 4  | 7   |
| ?  | ?   |

   d. Find a rule for your table in Question 2c.

3. What kind of algebraic expression do you think can be used as the rule for a linear function? (You might either give some examples or try to provide a general description.)

# *Using a Graph*

## Mathematical Topics

- Analyzing intuitively what type of expression defines a linear function
- Defining a linear expression
- Reviewing graphing on a graphing calculator
- Solving equations using graphing calculator graphs

## Outline of the Day

### In Class
1. Discuss *Homework 11: Line It Up*
2. Begin a review of graphing on a graphing calculator
3. *The Graph Solves the Problem*
   - Students use graphing calculator graphs to solve problems
   - No whole-class discussion of this activity is needed

### At Home
*Homework 12: Who's Right?*

## Discuss With Your Colleagues

### Can You Trust the Graphing Calculator?

In *The Graph Solves the Problem* and other activities in this unit, students are asked to solve an equation using a graphing calculator. Are answers that students get from calculator graphs reliable? What issues are raised by this use of technology?

# 1. Discussion of *Homework 11: Line It Up*

Have students compare their results within their groups on Questions 1a, b, and c. Unless there are disagreements, the whole-class discussion can probably begin with Question 1d. You can open the discussion by having several diamond card students explain their answers. The focus of the discussion should be on their justifications for their conclusions. (Presumably, they will see that the specific points from pairs in the table appear to lie on a straight line and will identify *f* as a linear function.) Bring out that Question 1d refers to the entire graph, and not just to the three specific points, so an explanation needs to refer in some way to the entire graph. *Do not expect a formal proof here,* but rather just an intuitive argument.

For example, a presenter might give a few more points from the table or graph and say, "As you move one unit to the right each time, you always go up three units" or "As *x* goes up by 1, *f(x)* will always go up by 3." Students also might use a phrase like "constant difference" to describe what's happening.

*Note:* Formally, the "constant difference" idea is related to "straightness" through the concept of similarity, but don't get distracted by looking for a formal explanation here.

On Question 2, let several students give the points they found in Question 2b and the rule they found for Question 2d, which should be equivalent to $y = 2x - 1$. If students all used the same point for Question 2b, ask for volunteers to come up with other points on the line through $(2, 3)$ and $(4, 7)$. Then discuss how the rule works for all the points on the line.

## • Question 3

Finally, let two or three volunteers share ideas about Question 3. They will probably state something to the effect that the expression should be a multiple of *x* plus some number. Tell students that this is correct and that an expression of the form $ax + b$ (where *a* and *b* are any numbers) is called a **linear expression in *x*,** because such expressions are the basis for defining linear functions.

Bring out that the term *linear* thus has two meanings—one geometric and one algebraic. You can point out that the term gets its algebraic meaning from the connection between the expression and the graph of the function it defines.

**"What are some other linear expressions in x?"**

**"What does it mean about the line if a is negative?"**

Ask students to give you a few more sample linear expressions, and have students graph several of the functions defined by these expressions on graphing calculators to see that they do give linear graphs. If someone gives you an example in which the coefficient $a$ is negative, ask what this means about the line. Students should see that the line will go down to the right instead of up to the right. (If no one gives you an example in which $a$ is negative, ask whether this is possible and, if so, what it means in terms of the graph.)

**"What's an example of a linear expression in w?"**

Also ask for some examples of linear expressions using variables other than $x$, so that students don't get stuck on the idea that the letter $x$ has a magical role to play.

**"What are some examples of nonlinear expressions?"**

Lastly, be sure to get some examples of expressions that are *not* linear in order to clarify the concept. As a hint, you can ask for an expression that defines a function whose graph is not a straight line. Students should be able to provide some simple examples such as $x^2$.

*Note:* We are dealing here only with linear expressions in one variable. Students will see later in the unit (*Homework 23: From One Variable to Two*) that the concept of a linear expression generalizes to several variables. However, linear equations (and inequalities) in two variables will be discussed more fully in the Year 2 unit *Cookies*.

## 2. Review of Graphing on the Graphing Calculator

In *Back to the Lake* (on Day 10), students worked with pencil-and-paper graphs, plotting individual points from an equation or an In-Out table. In today's activity, *The Graph Solves the Problem,* they will be using graphing calculators to get graphs more easily and accurately; they will then use these graphs to solve simple equations. In the last part of the unit, they will use calculator graphs to solve more difficult equations.

Ask students to briefly summarize their work on *Back to the Lake.* Their summary should include the In-Out table they created, the rule they found for this table expressing $d$ in terms of $N$ (probably the equation $d = \frac{6}{N}$, and the graph they made by plotting points from their In-Out table.

**"How else can you get a graph besides plotting individual points?"**

Ask what other method they have available for making graphs. If necessary, remind them of their past work making graphs of functions on the graphing calculator. Then have them make a graph on the graphing calculator of the function from *Back to the Lake,* defined by the equation $d = \frac{6}{N}$. They will be immediately confronted by the issue of change of variables, because the calculator will probably require the use of the variables $X$ and $Y$.

As needed, review graphing calculator mechanics, such as entering functions (and changing variables as needed), adjusting the viewing window, and using the trace feature.

> The standard viewing window will probably be good for this function, because most of the "action" involves small positive values for the two variables. You can ask students to find something like $f(20)$ on the calculator graph to force them to adjust the viewing window. (They will need to be able to adjust the viewing window in today's activity, *The Graph Solves the Problem*.)

Have students use the trace feature to see that points from their In-Out table are on the graph. *Caution:* They will probably only get points close to those in their table. For example, instead of finding (4, 1.5), they may come up with something like (4.11, 1.46).

- *Optional: Using a graph for "Catching Up"*

  If you think another example would be helpful here, you can use the situation from the Day 9 activity *Catching Up.* As needed, remind students that the Sawyers were traveling 30 miles per day (10 hours per day at 3 miles per hour) while the main group, which was 120 miles ahead, was traveling at 24 miles per day (8 hours per day at 3 miles per hour).

  Students probably got an equation like $120 + 24N = 30N$ to describe how many days it would take for the Sawyers to catch up. Now, you can have students graph two functions, defined by the equations $y = 120 + 24x$ and $y = 30x,$ and see where they intersect as a way to solve the problem.

## 3. The Graph Solves the Problem

After briefly reviewing graphing on the graphing calculator, have students begin work on *The Graph Solves the Problem*. In this activity, students will actually use the graph to find solutions to problems, rather than just confirm solutions as they did previously.

*Possible hint: "Does $277 (in Question 1a) represent a vertical coordinate or a horizontal coordinate?"*

One aspect of this task will be for students to decide whether the specific number they are given in the problem (such as 277 in Question 1a, or 6 in Question 2b) represents a vertical or a horizontal coordinate. This is further complicated by the need to change variables to $X$ and $Y$ in order to enter the functions on the graphing calculator. If groups are confused, it may help if you ask them a question directly on this issue, such as whether a particular number represents a vertical coordinate or a horizontal coordinate.

- *Two approaches*

  There are at least two distinct techniques that students might use with the graphing calculator to answer the questions in this activity. To illustrate, consider Question 1a, assuming that students are using the equation $p = 7t - 500$ (as described in the discussion of *Homework 10: What Will It Answer?*).

Students essentially are looking for the value of $t$ that corresponds to $p = 277$. With a change of variables to $X$ and $Y$, this means finding the value of $X$ that corresponds to $Y = 277$. Students might use either of these methods:

- Graph $Y = 7X - 500$ and use the trace feature to find the $X$-coordinate of the point on the graph whose $Y$-coordinate is 277.

- Graph both $Y = 7X - 500$ and $Y = 277$ and find the $X$-coordinate for the point of intersection.

Both of these approaches are fine, and you might look for opportunities to get students to describe each of them.

In effect, students are solving the equation $7X - 500 = 277$. Seeing the second approach might suggest to them what to do if they need to solve an equation in which the right side is more complicated than a single number.

### • Adjusting the viewing rectangle

The graphing calculator's standard viewing window will probably not show the portion of the function that students need on these problems, so they will have to make adjustments. We recommend that you give them plenty of opportunity to work through this issue in their groups, rather than simply give them general guidelines about how to set the window.

### • Approximation issues

There might be some questions about the degree of accuracy needed. For example, in Question 1a, students will be looking for a point on the graph of the equation $Y = 7X - 500$ whose $Y$-coordinate is 277, but they may be unable to obtain 277 as the exact value. As needed, discuss this with individual groups or with the entire class. There are several points to bring out.

- Answers obtained from the graphing calculator are likely to be approximations.

- An approximate answer may be satisfactory, but the degree of accuracy needed depends on the problem context.

- Zooming in generally yields as much accuracy as needed.

- The best way to test an estimated solution obtained from a graph is to go back to the equation itself.

## Homework 12: Who's Right?

Over the next few days, students will be building intuitions that will serve as the foundation for the distributive property, which is formally introduced on Day 15. Tonight's homework assignment is a lead-in to that work.

In particular, Question 1 introduces the concept of *equivalent expressions,* and Question 2 taps the connection made earlier in the unit between area and multiplication.

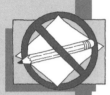

# The Graph Solves the Problem

In Question 2 of *Homework 10: What Will It Answer?* you probably came up with an equation like $p = 7t - 500$ to describe the theater's profit ($p$) in terms of the number of tickets they sold ($t$).

In Question 3 of that assignment, you were given the equation $d = 16t^2$ to describe $d$, the distance an object has fallen (in feet), in terms of $t$, the time elapsed (in seconds).

Use these two equations and the graphing feature of your graphing calculator to answer these questions.

1. a. The theater made a profit of $277 on yesterday's show. How many tickets were sold?

   b. Three hundred people bought tickets for today's show. How much profit did the theater make?

2. a. An object is dropped off the roof of a very tall building. How long will it take for the object to fall 200 feet?

   b. How far will the object have gone if it falls for 6 seconds?

*Note:* You may find that you want to use a method other than graphing to answer these questions. If so, use that method to *check* your answer *after* you have used the graph.

# Homework 12      Who's Right?

1. Andrew and Gladys were working on rules for In-Out tables. When they got to the table shown here, Andrew said that the *Out* at the bottom should be $2(X + 1)$. Gladys said it should be $2X + 2$. You need to decide who is right. Study the table and think about other In-Out pairs of numbers that you think would fit the pattern.

   | In | Out |
   |----|-----|
   | 3  | 8   |
   | 5  | 12  |
   | 9  | 20  |
   | 15 | 32  |
   | X  | ?   |

   If you think that either Andrew or Gladys is wrong, or that both are wrong, explain why you think so. If you think that they are both right, explain how there could be two different answers.

2. Find the area of each of the shapes shown here. That is, find out how many 1-foot–by–1-foot squares will fit in each without overlapping. Assume that all angles are right angles. *Do not assume* that these drawings are to scale.

   a.
   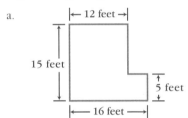

   b.
   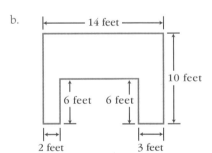

**Days
13–21**

# *What's the Same?*

*This page in the
student book
introduces Days 13
through 21.*

In everyday language, we can usually say the same thing
in many different ways. This is also true about the
language of algebra. Algebraic expressions that say the
same thing are called "equivalent."

In mathematical work, you often need to be able to switch
smoothly from one algebraic expression to an equivalent
one. In the next section of this unit, you'll be looking at
ways to do this and at how to use equivalent expressions
to solve equations. You may be surprised to see that you
can often use geometry to find equivalent algebraic
expressions.

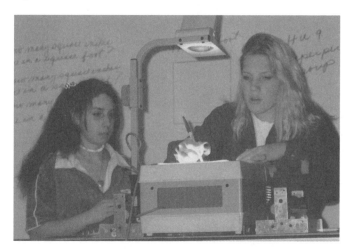

*Sofia Anis and Molly Berglund take delight in challenging the class to
solve some of the equations they have scrambled.*

# A Lot About Lots

*Students work with area as a model for multiplication.*

## Mathematical Topics

- Introducing equivalent expressions
- Using an area model for multiplication of algebraic expressions

## Outline of the Day

### In Class

1. Discuss *Homework 12: Who's Right?*

   - Introduce the concept of equivalent expressions
   - Review the relationship between area and multiplication

2. *A Lot of Changing Sides*

   - Students use area as a model for working with multiplication of simple algebraic expressions
   - Use Question 1 as an example to explain the activity

- The activity will be discussed on Day 14

### At Home

*Homework 13: Why Are They Equivalent?*

## Discuss With Your Colleagues

### What's the Big Deal About the Distributive Property?

The distributive property plays a major role in this unit. Why is the distributive property given such importance? Why are several ways provided for understanding it? And why isn't the property formally stated until students have dealt with the ideas for several days?

# 1. Discussion of *Homework 12: Who's Right?*

*Reminder:* Although Questions 1 and 2 of the homework may seem unconnected, both will be used in the process of building toward the distributive property.

You can have groups compare answers for the areas of the two figures in Question 2 while you check off which students did their homework.

- ## Question 1

  Students will probably recognize that the two rules, $2(X + 1)$ and $2X + 2$, are both correct and that they always give the same results. If any students argue otherwise, ask for a *counterexample* for whichever rule they don't like—that is, an In-Out pair of numbers that fits one rule but not the other. Make sure they realize that both $2(X + 1)$ and $2X + 2$ work for *all* the rows and that these two rules seem to give the same results *no matter what value is substituted for X.*

  Introduce the term **equivalent expressions** for two algebraic expressions that *always* give the same results. Tell students that this is an important concept that they will be applying throughout the rest of this unit and beyond. Tell them also that they will learn some powerful techniques for writing complicated algebraic expressions as equivalent simpler ones.

- ## Why are they equivalent?

  Point out to students that they can't possibly check all possible values for $X$ to confirm that $2(X + 1)$ and $2X + 2$ *always* give the same result. Inform them that in their homework tonight, they will be looking at and evaluating three proposed explanations of why these expressions are equivalent. (Because that is tonight's assignment, you should postpone discussion of the issue for now.) Also tell them that one of these explanations involves area and that the connection between area and equivalent expressions is the reason Question 2 was included in last night's homework.

- ## "Equivalent" means "equal"

  Tell students that although the expressions $2(X + 1)$ and $2X + 2$ are being called "equivalent," this really means that they give the same result, and that sometimes we say that they are *equal* rather than that they are *equivalent*. In other words, equivalent expressions are expressions that may look different but are actually equal to each other. You can also point out that the same language is used for fractions. For example, we call $\frac{1}{3}$ and $\frac{2}{6}$ "equivalent fractions," but we mean that they are equal.

## • *Equivalence and graphs*

To reinforce the relationship between an In-Out table and a graph, you can ask the class how the equivalence of the two expressions $2(X + 1)$ and $2X + 2$ might show up in terms of graphs. Students should realize that having the same In-Out table is really the same as having the same graph. To illustrate this, have students graph the functions defined by the two expressions $2(X + 1)$ and $2X + 2$ on their graphing calculators. They should see just one graph.

*Technical note:* You might want to have students first graph each function separately to confirm that they have the right viewing window to see the graphs and then graph them simultaneously to see that the graphs coincide.

## • *Focusing on symbolic expressions*

You may want to alert students that the unit will focus for a while on more formal work with algebraic expressions. The activities will still generally have some contextual setting. However, these contexts will be somewhat artificial and will serve more as models for understanding the algebra than as interesting problems in themselves.

Emphasize to students that algebraic expressions can be important tools in working with interesting problem situations. As those situations get more complicated, however, students need to be able to simplify algebraic expressions and replace them with equivalent ones.

## • *Question 2*

**"How did you find the area of each figure in Question 2?"**

You can have club card students from two groups explain how they found the area of each figure. Students will probably have subdivided the overall area into rectangles and added the areas of the pieces. To do so, they will need to find some dimensions that aren't given in the original diagram. Be sure that they justify this.

For example, in Question 2a, they might find that the side labeled "*X* feet" in the accompanying diagram must be 10 feet long because it combines with the 5-foot side to total 15 feet.

Then, if they use the dotted line shown here, they will get one rectangle that is 12 feet by 10 feet and another that is 16 feet by 5 feet. *Note:* There are other ways to subdivide this figure.

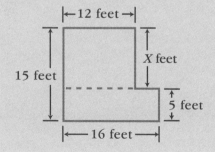

*"Why can you simply multiply length by width to get the area?"*

The main focus of the discussion should be on how students find the areas of the individual rectangles, because the purpose of this activity is to get them to understand the relationship between areas of rectangles and multiplication. They will probably automatically multiply length by width to find the area. Make them justify this shortcut, by bringing them back to the definition of area: "How many 1-foot–by–1-foot squares will fit inside without overlapping?"

In connection with this, as needed, remind students of the discussion of *Homework 2: Building a Foundation*. They should be able to articulate that an *m*-by-*n* rectangle can be thought of as having *m* rows with *n* unit squares in each row, so that its area is the sum $n + n + \ldots + n$, with *m* terms, that is, $m \cdot n$. (You might again point out that this is the rationale for the notation $m \times n$ for an *m*-by-*n* rectangle.)

> *Note:* Students may want to do Question 2b by subtracting a $9 \times 6$ rectangle from the overall $14 \times 10$ rectangle. This approach is fine.

## 2. *A Lot of Changing Sides*

> This activity is designed to give students more exposure to an area model of the distributive property and to a model for multiplying binomials. The activity will be discussed tomorrow.

### • *Introducing the activity*

*"What is the activity asking you to do?"*

Have students read the introduction and Question 1 of the activity, or perhaps have a volunteer read this material aloud to the class. Then ask students to explain what exactly they are being asked to do. Be sure that they identify the two steps of drawing a sketch and finding an expression for the new area.

To get them started, you can ask how they could sketch and label the original lot. They should get something simple like this:

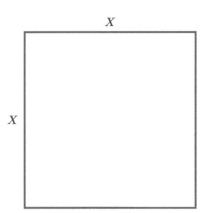

Then have them discuss in their groups how to sketch and label the lot described in Question 1. You may need to clarify what is meant by "in one direction" and "in the other." (Students may think this means *opposite* directions—that is, left and right—instead of *perpendicular* directions.) They should come up with something like this diagram in which the shaded area represents the "new" portion of the lot.

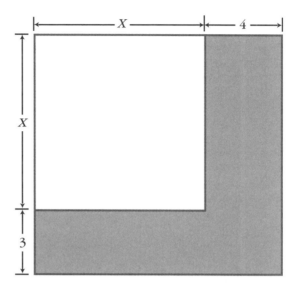

At this point, ask the groups to come up with two expressions for the total area as described in the activity, and have volunteers share their results. Students should see that the area is simply the product $(X + 3)(X + 4)$. If they need a hint to find another expression, suggest that they subdivide the area into rectangles, as on last night's homework. One way to do this is shown here, with the areas of the individual rectangles indicated. Thus, the area is $X^2 + 4X + 3X + 12$.

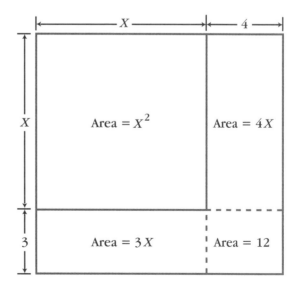

*Note:* Other answers are also possible based on variations in the subdivision of the diagram. Students do not need to combine terms, but it's okay if they do.

It is important that the expression without parentheses come from the diagram as a sum of areas, even if some students know how to multiply out the product $(X + 3)(X + 4)$. The ultimate purpose of the activity is to use the area analysis as an explanation for the multiplication process.

With the completion of Question 1 as a whole class, let students work in groups on the rest of the activity. *Note:* Question 5 is a challenge problem for groups that finish early, and groups do not have to complete it.

As students work on this activity, you can let them know that some of the earliest uses of area models to understand the multiplication of expressions like these was done by Islamic mathematician and astronomer al-Khwarizmi (ca. 780–850) and his followers.

## Homework 13: Why Are They Equivalent?

Tonight's homework assignment is a follow-up to today's discussion of last night's homework. This assignment should provide students with different ways to see why the expressions must be equivalent. Although the homework touches on the issue of proof, the focus is on building intuitive understanding.

*Kristin Livingston and Katy Anderson together prepare the overhead for their presentation.*

# A Lot of Changing Sides

A housing developer submitted plans to the city planner for some houses she wanted to build. The lots in the plan were all squares of the same size. But the city planner thought that this plan was boring and insisted that the developer introduce some variety. After some discussion, the planner and the developer decided that the lots should include other types of rectangles. So the developer proceeded to change the lengths of some of the sides of the lots.

For each of the changes that were made to the square lots, complete these tasks.

- Make and label a sketch of the lot, using the variable $X$ to represent the length of a side of the original square.

- Write an expression for the area of the new lot as a product of its length and width.

- Write an expression *without parentheses* for the area of the new lot as a sum of smaller areas. Use your sketch to explain this expression.

  1. The original square lot was extended 4 meters in one direction and 3 meters in the other.

  2. The original square lot was extended 5 meters in one direction only.

  3. The original square lot was extended 10 meters in one direction and 9 meters in the other.

  4. The original square lot was extended 1 meter in one direction and 25 meters in the other.

  5. The original square lot was extended 2 meters in one direction and decreased 2 meters in the other.

# Homework 13    Why Are They Equivalent?

You saw in *Homework 12: Who's Right?* that the two expressions $2(X + 1)$ and $2X + 2$ seem to give the same result no matter what number is substituted for $X$. In other words, the expressions appear to be equivalent. But it would be nice to be certain of this and to understand *why* the expressions are equivalent.

Randy, Sandy, and Dandy were having just that discussion. Read each of their explanations, and then do these four things.

1. Decide whether any, all, or just some of them are correct, and explain your decision.

2. State which explanation is the easiest for you to understand, and why.

3. State which explanation is most convincing to you, and why.

4. Adapt the explanation you understand best to explain in your own words why the expressions $3(X + 4)$ and $3X + 12$ are equivalent.

## *Randy's Explanation*

"We all know that $2A$ is twice $A$, which is $A + A$. Think of $2(X + 1)$ as being twice $X + 1$. In other words, it is equal to $(X + 1) + (X + 1)$. And $(X + 1) + (X + 1)$ is equal to $2X + 2$."

*Continued on next page*

## *Sandy's Explanation*

"It works with numbers! Check it out! If $X$ is 5, then $2(X + 1)$ is $2(5 + 1)$, which is the same as $2 \cdot 6$, which is 12. And, well, $2X + 2$ is $2 \cdot 5 + 2$, which is the same as $10 + 2$, which is also 12! Wow!"

## *Dandy's Explanation*

"Multiplication is how you find the area of a rectangle, you know, length times width. Basically, a product $ab$ can be thought of as the area of a rectangle with dimensions $a$ and $b$, like this:

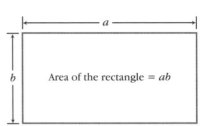

"The product $2(X + 1)$ can represent the area of a rectangle that is 2 units in one dimension and $X + 1$ units in the other. The length of $X + 1$ is like a segment of length $X$ next to a segment of length 1. The picture is something like this:

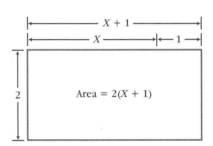

"A simple dividing line shows that this figure can be thought of as two rectangles, with areas $2X$ and 2, put together. Because we are talking about the same area, $2(X + 1)$ must equal $2X + 2$."

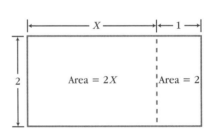

# Lots More

## Mathematical Topics

*Students use the area model for multiplication to explain equivalence of algebraic expressions.*

- Using equivalent expressions
- Showing equivalence of expressions in several ways
- Using an area model for multiplication of algebraic expressions

## Outline of the Day

### In Class

1. Discuss *Homework 13: Why Are They Equivalent?*

   - Be sure to get a good explanation of Dandy's area model

2. Discuss *A Lot of Changing Sides* (from Day 13)

   - Try to get a variety of diagrams

### At Home

*Homework 14: One Each Way*

## 1. Discussion of *Homework 13: Why Are They Equivalent?*

*"Whose argument is the most convincing?"*

*"Does Sandy's argument prove that the expressions are equivalent?"*

You can poll students to see if there is agreement on who is correct and ask for volunteers to restate the arguments of Randy, Sandy, and Dandy. (Tonight's homework refers to Randy's way as *repeated addition,* Sandy's way as a *numerical example,* and Dandy's way as an *area model.* Use these terms as you identify each method of explanation.)

All three arguments make valid points, although Sandy seems to be arguing with a single numerical example, which is not as convincing as the others' more general arguments. If no one brings this up, do so yourself and remind students that even doing many examples does not constitute a proof. Keep in mind, however, that the assignment is not really about proof, but about understanding why $2(X + 1)$ must equal $2X + 2$. Numerical examples *are* very helpful in this respect because they are easily understood and are based on something with which students are fairly comfortable, namely, arithmetic with whole numbers.

Because the area model is central to the development of the distributive property, be sure to discuss it. If no one volunteers to explain it, then give groups a little more time to read Dandy's explanation again, and then go over it with the class.

### • *Generalizing the conclusion*

*"What would be equivalent to 3(X + 7)?"*

Ask students to find an expression that is equivalent to $3(X + 7)$. Most likely, they will see that the proper expression is $3X + 21$. (If necessary, have them use their preferred method from the homework to find the expression. Sandy's method isn't really helpful in generating an expression, because all Sandy did was verify the equivalence with an example.) The key is for students to see that the constant term, 21, is obtained by multiplying the constant inside the parentheses, which is 7, by the factor outside, which is 3.

In preparation for tonight's homework, you should also do an example with subtraction, such as $5(X - 4)$. Students might use Randy's approach to see that this is equivalent to $5X - 20$.

Finally, ask for an equivalent to $a(X + b)$. Students should see that this seems to be equivalent to $aX + ab$ (although they might not be able to give an explanation for this general case). What's most important is that they realize that $a(X + b)$ is equivalent to $aX + ab$, and *not* to $aX + b$.

## 2. Discussion of *A Lot of Changing Sides*

You can assign one or two groups to each problem to present to the whole class. Because Question 5 is complicated by having a *decrease* in one direction, you may want to ask for volunteers to present it. As students go through their presentations, have them compare the area expressed as a sum of separate rectangles with the product of the new length and width.

For example, as discussed yesterday, the area in Question 1 might be expressed, using the diagram below, as $X^2 + 4X + 3X + 12$. On the other hand, the product of the length and width is $(X + 3)(X + 4)$.

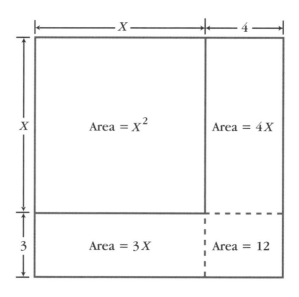

As review, ask what the term is for a pair of algebraic expressions, such as $X^2 + 4X + 3X + 12$ and $(X + 3)(X + 4)$, that represent the same thing. If necessary, remind students that these are called *equivalent expressions*.

You can have the class check the equivalence by substituting numerical values. Students can also check by graphing the functions defined by the equations $Y = X^2 + 4X + 3X + 12$ and $Y = (X + 3)(X + 4)$. They should see that the graphs are identical. (Be sure they know that neither of these "checks" proves the equivalence.)

### • *Combining terms*

**"How can you simplify the expression $X^2 + 4X + 3X + 12$?"**

If students haven't already mentioned it, bring up the idea of simplifying $X^2 + 4X + 3X + 12$. You should ask not only how the expression might be simplified but also ask what justification there is for combining $4X + 3X$ as $7X$.

Students might go back to their work earlier in the unit with the mystery bags, explaining that "four of something" plus "three of the same thing" is "seven of those things." Students might also use any of the three approaches from *Homework 13: Why Are They Equivalent?* to see why $4X + 3X$ is equivalent to $7X$. You might ask for volunteers to explain the equivalence for each method.

- In Randy's repeated addition method, $4X$ is four $X$'s added together and $3X$ is three $X$'s added together, so $4X + 3X$ is seven $X$'s added together, which is $7X$.

- In Sandy's numerical example method, students can substitute values for $X$ to see that $4X + 3X$ gives the same result as $7X$.

- In Dandy's area model method, students might use this diagram to show the same area as both $4X + 3X$ and as $7X$.

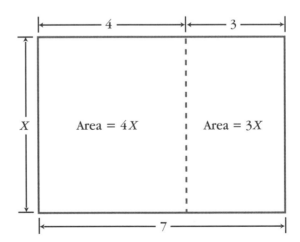

Whatever approach students use, tell them that this type of simplification is called **combining like terms.**

• *Question 5*

The area model falls apart somewhat when it comes to dealing with negative numbers, but students may have some creative ways of dealing with "negative area." Most students are comfortable with a diagram like the one here, even though it shows both negative lengths and negative areas. Although such a diagram is a bit unorthodox, you should encourage students to use it if it helps them work with the ideas, perhaps reminding them that length and area are not negative.

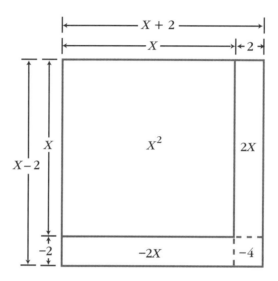

• *If time allows, do more examples*

If time permits, you may want to do some more examples, like those in *A Lot of Changing Sides*.

## Homework 14: One Each Way

This is a follow-up to today's discussion of *Homework 13: Why Are They Equivalent?*

# Homework 14 One Each Way

1. Find an equivalent expression without parentheses for each of these expressions.

   a. $5(A + 7)$

   b. $8(y - 4)$

   c. $2(W + 6)$

2. In *Homework 13: Why Are They Equivalent?* you saw three ways of thinking about why $2(X + 1)$ is equivalent to $2X + 2$. Now use those ideas to explain your work in Question 1.

   a. Use Randy's *repeated addition* method to explain why your answer to Question 1a is equivalent to $5(A + 7)$.

   b. Use Sandy's *numerical example* method to explain why your answer to Question 1b is equivalent to $8(y - 4)$.

   c. Use Dandy's *area model* method to explain why your answer to Question 1c is equivalent to $2(W + 6)$.

3. Find an equivalent expression without parentheses for each of these expressions. Use any method you like, but explain your work.

   a. $(r + 4)(r + 3)$

   b. $(3t + 1)(t + 5)$

# DAY 15 *Distributing the Area*

*Students formally state the distributive property and explain it using the area model.*

## Mathematical Topics

- Expressing the area of a rectangle both as the product of its length and width and as a sum of smaller areas
- Making a formal statement of the distributive property
- Using the area model to understand the distributive property
- Using the distributive property for both factoring and multiplying

## Outline of the Day

### In Class

1. Select presenters for tomorrow's discussion of *POW 2: Tying the Knots*
2. Discuss *Homework 14: One Each Way*
   - Discussion of this homework might be omitted
3. *Distributing the Area*
   - Students express different areas algebraically
4. Discuss *Distributing the Area*
   - Focus on how the total is the algebraic sum of its parts

5. State the distributive property formally
   - Post both a statement of the property and an area explanation
   - Discuss factoring and multiplying as two aspects of the property

### At Home

*Homework 15: The Distributive Property and Mystery Lots*

# 1. POW Presentation Preparation

Presentations of *POW 2: Tying the Knots* are scheduled for tomorrow. Choose three students to make POW presentations, and give them overhead transparencies and pens to take home for preparing presentations.

# 2. Discussion of *Homework 14: One Each Way*

If students seem comfortable with these ideas, you may want to omit discussion of most of this assignment. Because Question 3b is a bit more complicated than earlier "lots" examples, however, you might have a student present this problem. In particular, go over the fact that $3t \cdot t = 3t^2$.

You may want to collect this assignment to assess students' understanding of the concepts.

# 3. *Distributing the Area*

This activity gives students one more look at the area model for understanding the distributive property before stating the property formally.

The activity is quite structured, at least through the various parts of Question 1, so students should be able to move fairly easily through most of it.

If groups need a hint on Question 4, you can ask what the sides of the rectangle should be if its area is the product of $p + q + r$ and $x + y + z$.

*"What is like area but involves three factors?"*

For groups that get to Question 5, you might give an additional hint such as asking, "What is like area but involves three factors?"

When most groups have finished Question 4, you can bring the class together to go over the individual problems.

# 4. Discussion of *Distributing the Area*

Questions 1a through 1d of the activity are fairly straightforward, so you might start the discussion with Question 2. You can ask for volunteers to share their answers with the class, making sure you get both types of expressions called for in the problem.

In Question 3, students should suggest at least these two approaches (and perhaps others) for expressing the area:

- As a product of length and width: $(a + b)(c + d)$
- As a sum of smaller areas: $ac + ad + bc + bd$

Then move on to Question 4. Someone should be able to come up with the idea of a rectangle with sides $p + q + r$ and $x + y + z$ and with component subrectangles of areas $px, py, pz, qx, qy, qz, rx, ry,$ and $rz$. Thus:

$$(p + q + r)(x + y + z) = px + py + pz + qx + qy + qz + rx + ry + rz$$

Question 5 can be left unresolved if no one has any ideas. (You can return to it later in the year when students study volume in *Do Bees Build It Best?*)

## 5. The Distributive Property: Officially!

Tell students that the area model they have been working with is a way of understanding a deep and pervasive principle of algebra, known as the **distributive property** (also called the *distributive law*). Basically, they will be making a transition from this geometric model to a more symbolic representation of the same idea.

Tell students that the distributive property is probably the most frequently used principle for working with equivalent expressions and that they need to be aware of this principle in many guises. You can either ask students if they know what the distributive property is or just tell them.

The simplest form of the distributive property might be expressed by an equation such as $a(x + y) = ax + ay$. You should post a statement of the distributive property, leaving room for a diagram.

*"Can you state the distributive property in words?"*

Ask for a volunteer to put this equation into words. For example, a student might say,

> "Multiplying a sum by some factor is the same as multiplying each term of the sum by that factor and then adding the products."

*"What kind of diagram should we put with the equation $a(x + y) = ax + ay$ to explain the distributive property?"*

It's a good idea to have students make an area diagram that illustrates this, just to be sure that they can relate the symbols to the area. (This repeats earlier work, but this connection really can't be overdone.) They should be able to come up with a diagram like this one, which you can post along with the statement of the distributive property.

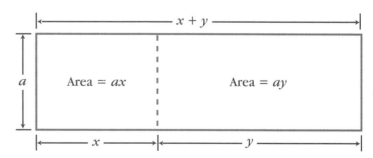

**Area of large rectangle = $a(x + y)$**
**Sum of areas of small rectangles = $ax + ay$**
$a(x + y) = ax + ay$

## • *Different ways of looking at the distributive property*

There are actually several ways of looking at the distributive property.

**Factoring versus multiplying:** Point out that the equation $ax + ay = a(x + y)$ can be viewed as having two directions.

- Going from $ax + ay$ to $a(x + y)$, which is called **factoring** (or *factoring out*)

- Going from $a(x + y)$ to $ax + ay$, which is just **multiplying** (or *multiplying through*)

(Other terms for these steps are mentioned in tonight's homework. You may want to review these as well.)

> *Note:* Although we are introducing the concept of factoring here, students get only a brief overview of the idea. They will see more about factoring in the Year 3 unit *Fireworks*.

Tell students that in solving equations, sometimes one needs to factor expressions, while at other times one needs to multiply them out. Thus, students must be able to use the distributive property in both directions. You may want to have students do a few very simple examples in each direction, such as factoring $3x + 6y$ as $3(x + 2y)$ or multiplying $x(4x + 3)$ as $4x^2 + 3x$.

**Left versus right:** The distributive property also has both a "left" and a "right" form.

- Left distributive property:   $a(x + y) = ax + ay$

- Right distributive property: $(x + y)a = xa + ya$

Although this distinction may seem minor, it should be pointed out. Otherwise, you may end up with students who are comfortable rewriting $5(a + b)$ as $5a + 5b$ but are stuck when they see $(x + 2)y$.

## Homework 15: The Distributive Property and Mystery Lots

> Part I of tonight's homework is intended to solidify the basic idea of the distributive property. Part II is essentially an introduction to factoring quadratic trinomials, but the goal is to use factoring as a vehicle for continued work with the distributive property. You may want to do the first example with the class as a review of the ideas in *A Lot of Changing Sides*.

# Distributing the Area

The figure on the right is a large rectangle made up of some smaller rectangles. The measurements of the smaller rectangles are shown using the variables *a*, *b*, *c*, and *d*.

Remember that you can compute the area of a rectangle by multiplying its length by its width.

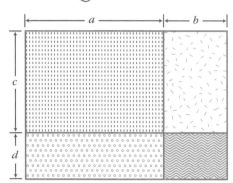

1. Use the "length times width" area formula and the variables *a*, *b*, *c*, and *d* to write an expression for the area of each of the smaller rectangles from the diagram.

   a. Area of the rectangle shaded like        = –?–

   b. Area of the rectangle shaded like        = –?–

   c. Area of the rectangle shaded like        = –?–

   d. Area of the rectangle shaded like        = –?–

*Continued on next page*

2. Next, look at certain combinations of rectangles and write each area in two ways.

   • As the product of its length and width

   • As the sum of two smaller areas

   a.  Area of the figure shaded like  = –?–

   b.  Area of the figure shaded like  = –?–

3. Write the area of the entire rectangle in *at least* two ways.

4. a.  Draw and label a rectangle whose area can be written as the product $(p + q + r)(x + y + z)$.

   b.  Show how to use your diagram to write the product $(p + q + r)(x + y + z)$ as an expression without parentheses.

5. *Challenge:* Draw a diagram that could be used as a model for finding an expression without parentheses that is equivalent to $(a + b)^3$.
   (*Hint:* We can't call it an *area* model.)

# Homework 15    The Distributive Property and Mystery Lots

## Part I: The Distributive Property

In its simplest form, the distributive property says that the expressions $a(x + y)$ and $ax + ay$ are equivalent. In other words, according to the distributive property, the equation

$$a(x + y) = ax + ay$$

is true no matter what numbers are substituted for $a$, $x$, and $y$.

Sometimes this property is used to replace an expression with parentheses by an equivalent expression without parentheses. For example, you can write $5(2x + 3)$ as $10x + 15$. This use of the distributive property is often called **multiplying through.** We also sometimes say that the factor 5 has been "distributed" across the sum $2x + 3$.

*Continued on next page*

Interactive Mathematics Program                                45

1. Distribute the factor across each sum.

   a. $4(a + 9)$

   b. $3(5w + 2r)$

   c. $6t(2 + 3s)$

   d. $10c(u + v + w)$

The distributive property is also used in the reverse direction from the examples of Question 1. For instance, you can write $8x + 12y$ as $4(2x + 3y)$. This use of the distributive property is often called **taking out a common factor** (or just *factoring*). You might want to think of this as "undistributing."

2. Take out a common factor in each of these sums.

   a. $14d + 21e$

   b. $rg + rh$

   c. $2pq + 4pr$

   d. $6ab + 10ac$

## Part II: Mystery Lots

The developer from *A Lot of Changing Sides* has taken to writing plans for lot sizes in algebraic code. Unfortunately, the codes are giving the city planner a hard time.

3. Here is what the city planner found written in the developer's notes one day:

   "Build a lot whose area is $X^2 + 3X + 5X + 15$."

   Help the city planner by finding out what the developer planned to do with the original square lot. (Remember that the original lot had sides of length $X$.)

4. What do you suppose each of these two entries means?

   a. "Build a lot whose area is $X^2 + 4X + 6X + 24$."

   b. "Build a lot whose area is $X^2 + 6X + 2X + 12$."

5. Then the developer's entries got even more cryptic. Figure out for the city planner what each of these entries means.

   a. "Build a lot whose area is $X^2 + 9X + 18$."

   b. "Build a lot whose area is $X^2 + 7X + 10$."

   c. "Build a lot whose area is $X^2 + 6X + 8$."

   d. "Build a lot whose area is $X^2 + 5X + 4$."

# POW 2 Presentations and Solidifying the Distributive Property

*Students present POW 2 and continue work with the distributive property.*

## Mathematical Topics

- Analyzing a probabilistic problem using a combinatorial approach
- Continuing work with the distributive property

## Outline of the Day

### In Class

1. Presentations of *POW 2: Tying the Knots*
   - Students should explain how they found the probability of each possible outcome

2. Discuss *Homework 15: The Distributive Property and Mystery Lots*

3. Discuss how ordinary multiplication of a two-digit number by a one-digit number illustrates the distributive property

### At Home

Homework 16: Views of the Distributive Property

# 1. Presentations of *POW 2: Tying the Knots*

Have the selected students give their presentations. Different students may have arrived at their solutions in different ways. They should find that the probability of one loop is $\frac{8}{15}$, of two loops is $\frac{6}{15}$, and of three loops is $\frac{1}{15}$.

> ### • *Teacher background: A possible explanation*
>
> Here is one possible approach to finding these probabilities. Number the strings from #1 through #6. We can assume that the upper knots have been tied and that #1 is tied to #2, #3 to #4, and #5 to #6. What matters now is how the lower knots are tied.
>
> There are five choices as to which string should be tied below to #1. Once this decision is made, pick one of the four remaining untied strings. There will be only three choices as to which string should be tied to this one. And once that decision is made, there are only two strings left, so they must be tied to each other. Thus, there are 15 possibilities.
>
> (*Note:* Students may not immediately see why the numbers of choices, 5, 3, and 1, are being multiplied here. It may help to actually list all the ways that the lower knots can be tied. Here is a way to represent them:
>
> > 1 to 2, 3 to 4, 5 to 6
> >
> > 1 to 2, 3 to 5, 4 to 6
> >
> > 1 to 2, 3 to 6, 4 to 5
> >
> > 1 to 3, 2 to 4, 5 to 6
> >
> > 1 to 3, 2 to 5, 4 to 6
> >
> > 1 to 3, 2 to 6, 4 to 5
>
> and so on.
>
> Students should see that for each of the five possible strings to tie to string #1, there are three ways to complete the knots, for a total of 5 • 3 ways.)
>
> Students still must figure out how many of these 15 possibilities give each configuration of loops. Of course, one approach is to test each case. But try to get students to be more analytical. Here is one method.
>
> The only way to get three separate loops is the first case listed above, in which the lower loops are tied exactly like the upper loops.
>
> To get a single large loop, string #1 must be tied to something other than string #2, so there are four choices as to which string is tied to #1. First consider the case in which #1 is tied to #3. Once this is done, #2 must be tied to #4, #5, or #6. But if #2 is tied to #4, then #5 and #6 must be tied to each other so they will form a separate loop. Thus, to get one large loop, #2 must be tied to either #5 or #6, which is two cases.
>
> Similarly, there are two cases in which #1 is tied to #4, two cases in which #1 is tied to #5, and two cases in which #1 is tied to #6. Thus, there are eight cases altogether that result in a single large loop.

The remaining six cases must result in one small loop and one medium loop.

Here is a way to explain independently why there are six cases involving a small loop and a medium loop.

To get a small loop, exactly one of the three pairs tied above must also be tied below. There are three choices: #1 and #2, or #3 and #4, or #5 and #6. If, for example, #1 is tied to #2, then #3 must be tied to something *other than* #4 in order to avoid having three separate loops. Thus, there are two choices for the

"partner" for #3. In other words, there are two cases in which #1 and #2 form a small loop and the other strings together form a medium loop.

Similarly, there are two cases in which #3 and #4 form a small loop and the other strings together form a medium loop, and two cases in which #5 and #6 form a small loop and the other strings together form a medium loop.

Thus, there are six cases altogether that result in one small loop and one medium loop.

- **"POW 3: Divisor Counting" to be introduced on Day 17**

The next POW involves finding a way to determine how many divisors an integer has without actually counting them. This POW will be introduced on Day 17 by an activity involving the concepts of prime numbers and prime factorization.

## 2. Discussion of Homework 15: The Distributive Property and Mystery Lots

You may simply want to let group members compare answers for Part I. If there are issues that groups cannot resolve, you can discuss them as a whole class.

For Part II, have volunteers present their answers. Insist that they explain how they got their answers. If you went over Question 3 in introducing the assignment, you can move on to Question 4.

For example, in Question 4a, students should see that they can get an additional $4X$ by extending one side by 4 meters. If they extend the other side by 6 meters, that will give both a $6X$ and a $4 \cdot 6$, which is exactly what is needed. Question 4b is similar.

For the problems in Question 5, students need to apply the idea that

$aX + bX = (a + b)X$ and decide how to split up the $X$ term of the expression to make the problem work. They may be able to articulate that they need two numbers whose sum is the coefficient of $X$ and whose product is the constant term, or they may simply do this process without explicitly identifying it that way. Either is okay.

Keep in mind that mastery of factoring quadratic trinomials is not the goal of this unit. We are more interested in a conceptual grasp of factoring and using the area model for multiplication as a foundation for understanding the distributive property.

Students will get some additional experience with factoring in *Fireworks* in Year 3. *Fireworks* contains supplemental problems that can be used by students who want to learn even more about factoring.

## 3. Ordinary Multiplication and the Distributive Property

Students often can better appreciate the distributive property by seeing its relationship to the ordinary multiplication of numbers. You can start by pointing out that many people use the distributive property when they do mental multiplication of a two digit number by a one-digit number. For example, to multiply 4 · 13, they might break down 13 into 10 + 3, multiply both 10 and 3 by 4, and then add up the products. This process will probably be clearer if it is written out as a series of equalities.

$$4 \cdot 13 = 4 \cdot (10 + 3)$$
$$= (4 \cdot 10) + (4 \cdot 3)$$
$$= 40 + 12$$
$$= 52$$

You can have students try some similar examples in their heads, such as 7 · 12 or 8 · 34. You might also include an example in which the *second* factor is a one-digit, number, such as 23 · 9. In either case, tell students that part of tonight's homework builds on this idea.

## Homework 16: Views of the Distributive Property

This assignment gives students some other ways to think about the distributive property, in terms of both familiar arithmetic and area diagrams,

and has them apply these ideas to the multiplication of algebraic expressions.

# Homework 16    Views of the Distributive Property

The **distributive property** is an important general principle that can be used in many situations to write a mathematical expression in another form. Recall that in its simplest algebraic form, the distributive property can be expressed by an equation like this one.

$$a(x + y) = ax + ay$$

In words, you might state the distributive property this way.

> Multiplying a sum by something is the same as multiplying each term by that "something" and then adding the products.

In this assignment, you'll be looking at various ways to think about and use the distributive property.

## *Multidigit Multiplication*

You may not have realized that you've been using the distributive property every time you do multiplication that involves more than one-digit numbers. For example, the product 73 · 56 can be thought of as (70 + 3) · 56. Applying the distributive property, you would get 70 · 56 + 3 · 56.

You might write this problem in vertical form.

$$
\begin{array}{r}
56 \\
\times\, 73 \\
\hline
168 \\
3920 \\
\hline
\end{array}
\quad
\begin{array}{l}
\\
\\
\text{(this is } 3 \cdot 56) \\
\text{(this is } 70 \cdot 56)
\end{array}
$$

*Comment:* People often omit the zero in 3920, simply multiplying 7 · 56 to get 392 and then writing 392 with the 2 lined up in the tens column.

*Continued on next page*

Each of the products 3 · 56 and 70 · 56 can also be found using the distributive property. To show all the details in the written multiplication, you might write it like this.

$$
\begin{array}{r}
56 \\
\times\,73 \\
\hline
18 \\
150 \\
420 \\
\underline{3500}
\end{array}
$$

     (this is 3 · 6)
     (this is 3 · 50)
     (this is 70 · 6)
     (this is 70 · 50)

Each of the numbers 18, 150, 420, and 3500 is called a **partial product.** Writing a multidigit multiplication showing all the partial products is sometimes called the **long form.**

In the usual written form of this problem, the partial products 18 and 150 are not shown individually. Instead, their sum, 168, is written. Similarly, we omit the partial products 420 and 3500 and simply write their sum, 3920. The numbers 168 and 3920, which are each the sum of two partial products, are sometimes referred to as **partial sums.**

1. Show how to find the product 32 · 94 using the long form, showing all the partial products.

## Multiplication with a Diagram

You can illustrate the product 73 · 56 with an area diagram like this one, in which each smaller rectangle represents one of the partial products. Notice that the areas of the two smaller rectangles on the right add up to the partial sum 168 while the areas of the two larger rectangles on the left add up to the partial sum 3920.

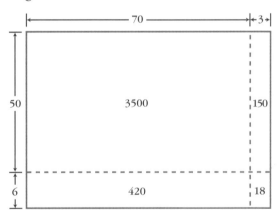

*Continued on next page*

2. Draw an area diagram like the one shown on the previous page to illustrate the product 32 · 94.

3. Show how to find the product 47 · 619 in two ways.

   a. Using the long form

   b. Using an area diagram

## Multiplying in Algebra Is Like Multiplying in Arithmetic

Multiplication of algebraic expressions can be done in a way that is similar to multidigit multiplication. For example, you can set up the problem $(x + 3)(2x + 5)$ in vertical form.

$$2x + 5$$
$$\underline{\times \ x + 3}$$

As with 73 · 56, this problem involves four separate products.

4. a. Find this product using a vertical multiplication form. You can use either the long form or a shorter form.

   b. Show how to do this problem using an area diagram.

# DAY 17

# *The Distributive Property and Divisor Counting*

*Students begin a POW about counting divisors.*

## Mathematical Topics

- Using the multiplication algorithm and an area model to understand multiplication of algebraic sums
- Working with prime numbers and prime factorizations

## Outline of the Day

### In Class

1. Discuss *Homework 16: Views of the Distributive Property*
   - Use the multiplication algorithm and an area model as tools for understanding multiplication of algebraic sums

2. *Prime Time*
   - Students find the prime factorizations of integers from 2 through 30

3. Discuss *Prime Time*
   - Bring out that every positive integer can be written as a product of powers of primes and tell students that this factorization is unique

4. Introduce *POW 3: Divisor Counting*
   - Suggest to students that they use their work on *Prime Time* for ideas

### At Home

*Homework 17: Exactly Three or Four*

*POW 3: Divisor Counting* (due Day 23)

# 1. Discussion of *Homework 16: Views of the Distributive Property*

You can have spade card members of different groups present Questions 1, 2, 3a, and 3b.

In discussing the long form of multiplication, you may need to review some basics of place value so students see where the 0's are coming from. You might also bring out that the standard multiplication algorithm (the "short form") involves combining some terms mentally. This is indicated in the homework but may be worth repeating.

## • *Question 4*

You may want to have several presenters for Question 4a, because students are likely to have different methods. Some students may use the long form. They may have various ideas about how (or whether) to line up the terms, because they don't have the usual place-value considerations to work with. For instance, they might produce either of these forms.

$$
\begin{array}{r}
2x + 5 \\
\times\ x + 3 \\
\hline
15 \\
6x \\
5x \\
2x^2 \\
\hline
\end{array}
\qquad \text{or} \qquad
\begin{array}{r}
2x + 5 \\
\times\ x + 3 \\
\hline
15 \\
6x \\
5x \\
2x^2 \\
\hline
\end{array}
$$

Other students may follow a pattern that looks more like the standard multiplication algorithm. For instance, students might produce either of these expressions.

$$
\begin{array}{r}
2x + 5 \\
\times\ x + 3 \\
\hline
6x + 15 \\
2x^2 + 5x \\
\hline
\end{array}
\qquad \text{or} \qquad
\begin{array}{r}
2x + 5 \\
\times\ x + 3 \\
\hline
6x + 15 \\
2x^2 + 5x \\
\hline
\end{array}
$$

There is no need to establish a single method for doing problems like this. You should emphasize that the purpose of a written format is to make the work as clear and as easy to do as possible, and so different students may prefer different methods.

After getting several approaches to Question 4a (preferably including at least one long-form method and one short-form method), have a volunteer present a diagram for Question 4b. The diagram should look something like this.

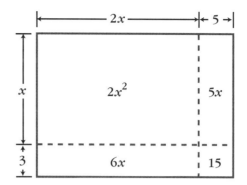

If needed, bring out the analogy between this diagram and the similar diagram in the assignment. Also point out the connection between the individual rectangles in the diagram and the partial products in the long form of the multiplication process.

### • *Combining terms*

Whatever method students use for getting the partial products or partial sums, they do need to know how to combine the terms into a single final expression. Thus, you will want to review the idea of combining like terms. You can use an analogy with place value, bringing out that multiples of $x$ are like 10's, multiples of $x^2$ are like 100's, and so on.

### • *Generalizing the distributive property*

*"In general, how do you multiply two sums together? What are the partial products?"*

Ask students to state a general principle for multiplying two sums together. As a hint, ask where the various partial products will come from. Students should be able to articulate that the partial products are all the products obtained by multiplying a term of one sum by a term of the other sum.

Have students explain this principle using an area diagram, and emphasize the value of being able to go back and forth between the symbol manipulation and the area model for multiplication. You want students to have a way to understand this multiplication process, rather than simply memorize a rule such as "FOIL" (which stands for *First, Outside, Inside, Last*). You may also want to have students work out an example in which each factor has more than two terms, such as $(x + y + 3)(2x + 3y + 1)$.

### • *The distributive property and multiplying negative numbers*

The distributive property can be used as a way of explaining the fact that the product of two negative numbers is positive. The supplemental problem *Preserve the Distributive Property* gives students an opportunity to explore this connection.

## 2. *Prime Time*

Students may have varied experiences with prime numbers. This activity provides a basic introduction or review, as the case may be, and is background for the next POW.

You will probably find it useful to work through an example of how to find the prime factorization. The number 60 provides a good example because it involves several primes and a repeated factor.

*"How can you write 60 as a product?"*

You can begin by asking for different ways to write 60 as a product, bringing out immediately that these examples show that 60 is not a prime number. Then have students break down the factors for one of these cases into smaller factors. Show them how to write this process as shown below, and introduce the term **factor tree** for this method of representing the process and the term **branch** for the splitting of each term into its factors. Bring out that one can "pick the fruit" off the "branches" of this "tree" to get the prime factorization, $2 \cdot 3 \cdot 2 \cdot 5$.

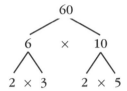

## 3. Discussion of *Prime Time*

You can have group members compare their lists of primes. If there are discrepancies that they can't resolve, have the whole class clear up any misconceptions.

You do not need to go over the prime factorization for all the composite numbers, but you should get the factorizations for a couple of examples so you can bring out that all students got the same factorizations (except perhaps for the order of factors).

Tell students that the prime factorization of a composite number is unique but that it is surprisingly difficult to prove this. (The proof requires an axiom about the set of natural numbers, such as the principle of induction.)

Discuss at least one example of the use of exponents to write the prime factorization. Present an example in which two distinct primes are used (other than the case of 12 shown in the assignment). For instance, you might do the number 18, which students should write as $2 \cdot 3^2$.

*"Can you write $2 \cdot 3^2$ so that both primes have exponents?"*

Ask if there is a way to write this so that both primes have exponents. Students should see that they can use an exponent of 1 with the prime 2 and thus write 18 as $2^1 \cdot 3^2$. Tell them that this expression is sometimes called the **prime power factorization.**

## 4. Introduction of *POW 3: Divisor Counting*

*"What would be the problem statement for this POW?"*

Have a volunteer read *POW 3: Divisor Counting* aloud. Then discuss with the class what the problem statement would be for this POW. Emphasize that the task is not to find out how many divisors any particular number has, but to find patterns in the numbers with a certain number of divisors. Students should explore divisors and see what they can find.

Be sure that students are clear about the definitions of *divisor* and *prime*. Emphasize that divisors here have to be whole numbers that "divide evenly" (no remainder). *Comment:* Students may see a resemblance between this POW and Question 2 of *Homework 2: Building a Foundation*.

This is an exploration POW; it differs from other POWs in that only a topic, and maybe some hints, is given to the students. *It is the students' responsibility to come up with questions and make conjectures.* Work on this type of POW usually begins with some information gathering, followed by the formulation of some questions and finally some conjecture about the way things are.

Exploration POWs are extremely difficult to do in one night, because finding patterns requires students to look at data in different ways. It is easy to get locked into looking at the data in one particular way, so students may need to take a break and then reexamine the data from a fresh perspective.

Be sure students see that the write-up categories for this POW differ significantly from those in the standard POW write-up.

This POW is scheduled to be discussed on Day 23.

## • *Homework gets students started*

Point out to students that their homework tonight will get them started on this problem. You can have groups use the rest of today's class time to work on the homework assignment and the POW.

You might tell students that their initial work will probably involve information gathering. Groups can begin by finding all the divisors of the different numbers and then combining these results into a table showing how many divisors each number has. You might suggest that students use their results from *Prime Time* to generate ideas for the patterns sought in Question 1b of the homework.

*Note: The Locker Problem* is a supplemental problem that involves mathematical ideas similar to those in this POW. You may want to use that supplemental problem as a follow-up or alternative to this POW.

## *Homework 17: Exactly Three or Four*

This assignment gives students a concrete way to get started on the POW.

Classwork

*Solve It!*

# Prime Time

A **prime number** (also called simply a *prime*) is a whole number that has exactly two whole-number divisors: 1 and itself. For example, 7 is a prime number, because it has exactly two whole-number divisors: 1 and 7. On the other hand, 10 is not a prime, because it has four whole-number divisors: 1, 2, 5, and 10. A nonzero whole number with more than two whole-number divisors is called a **composite number.**

The number 1 is *not* considered a prime, because it has only one whole-number divisor, nor is it a composite number. Rather, it is considered a special case.

1. Examine each of the numbers from 2 through 30.

   a. Make a list of the numbers in this group that are primes.

   b. For each composite number from 2 through 30, write that number as a product of prime numbers. This is called the **prime factorization.**
   *Note:* You may need to use more than two factors, and you can use the same factor more than once. For example, the prime factorization of 12 is the product 2 · 2 · 3.

2. If an expression from Question 1b used a factor more than once, rewrite the expression using an exponent instead of repeating the factor. For example, write 12 as $2^2 \cdot 3$ instead of 2 · 2 · 3.

# POW 3 *Divisor Counting*

As you saw in *Prime Time,* a prime number is a whole number that has exactly two whole-number divisors. This POW is about counting the divisors for any whole number—not only primes. Throughout this problem, the word *divisor* will mean *whole-number* divisor.

The number 1 is a divisor of every whole number and every whole number is a divisor of itself. Therefore, every whole number greater than 1 has at least two distinct divisors and so must either be a prime or have more than two divisors. Your task in this POW is to figure out as much as you can about *how many* divisors a number has. You will probably find the concept of prime numbers useful both in conducting your investigation and in stating your conclusions.

*Continued on next page*

Here are some examples of questions to look at.

- What kinds of numbers have *exactly* three divisors? *exactly* four? and so on.

- Do bigger numbers necessarily have more divisors?

- Is there a way to figure out how many divisors 1,000,000 (one million) has without actually listing and counting them? How about 1,000,000,000 (one billion)?

- What's the smallest number that has 20 divisors?

But you should not answer only these questions. You should also come up with your own questions and look for generalizations.

# Write-up

1. *Subject of Exploration:* What were you exploring? What were your goals?

2. *Questions:* What questions did you ask yourself? Why did you ask them? Which ones did you decide to explore?

3. *Information Gathering:* Based on your notes, describe what you did to get data for your exploration.

   a. How did you get started?

   b. What approaches did you try?

   c. What information did you gather?

   d. When did you decide to stop, and why?

4. *Results and Conjectures:* What conjectures did you find as possible answers to your questions? What rules or patterns did you find in exploring your questions? If you can prove that your conjectures are right, do so. If you can explain why a particular rule or pattern works, do that as well. If possible, generalize your results.

5. *Evaluation*

6. *Self-assessment*

# Homework 17 Exactly Three or Four

1. By definition, a number with exactly two whole-number divisors is a prime number. But what about numbers with exactly three whole-number divisors? Or exactly four?

   To get you started on *POW 3: Divisor Counting,* your first task in this assignment is to consider these two special cases.

   a. Find several numbers that each have exactly three divisors and several others that each have exactly four divisors.

   b. Examine your two lists and look for some explanations or patterns.

2. Decide on at least one specific question about divisors that you want to try to answer as part of your POW, and state that question as clearly as you can.

# DAY 18 Taking Some Out

*Students use the hot-and-cold-cube model to work with subtraction of sums.*

## Mathematical Topics

- Finding numbers with exactly three or exactly four divisors
- Understanding subtraction of sums in parentheses

## Outline of the Day

### In Class

1. Discuss *Homework 17: Exactly Three or Four*
   - Have students give examples of numbers with exactly three or exactly four divisors
2. *Taking Some Out, Part I*
   - Students use the hot-and-cold-cube model to work with subtraction of a sum

3. Discuss *Taking Some Out, Part I*
   - Formalize and post the algebraic rule for subtracting a sum

### At Home

*Homework 18: Subtracting Some Sums*

---

### Discuss With Your Colleagues

Parentheses and Subtraction

In traditional algebra courses, students often learn a rule such as "change the sign and add" for dealing with subtraction of expressions in parentheses. In today's activity, *Taking Some Out*, students are asked to think about this process using the hot-and-cold-cube model. Why is this model used here? What explanation do *you* like best for simplifying such expressions? How well has your favorite explanation worked with students in traditional classes?

---

## 1. Discussion of *Homework 17: Exactly Three or Four*

Perhaps all you need to do with Question 1 of this assignment is have students give examples of numbers with exactly three or exactly four divisors. As long as they can do this, they should be equipped to do further work on the POW on their own.

You can have students hold onto this assignment and turn in Question 2 as part of their POW write-ups.

## 2. Taking Some Out, Part I

This activity moves on from the distributive property to another aspect of simplifying algebraic expressions, namely, subtracting expressions in parentheses. This is a difficult topic for many students, and there is no "magic bullet" for making sense out of it. This activity refers to the hot-and-cold-cube metaphor developed in *Patterns* for dealing with positive and negative numbers, although it actually only uses hot cubes.

Read the introduction and the first problem together as a whole class and use it as an example. The idea is to write the situation both as 50 – (5 + 10) and as 50 – 5 – 10, and to point out that these two expressions represent the same number. (You may want to acknowledge that this entire activity is a bit contrived.)

Then have students work in groups on the other two problems. Question 2 should be fairly routine because it is similar to Question 1.

Question 3 gets more deeply at the idea of removing parentheses in a subtraction problem. This could be one of those problems for which some students think the answer is obvious—"What does it matter if you take the two groups out together or separately?"—while others struggle to understand what's going on. Be sure that students who get the point right away take the time to write the answer symbolically, because the symbol manipulation is what most troubles students.

As groups finish the graphing in Question 3c, you can assign them a part of the assignment to present to the class. You can also have students who finish quickly work on their POW and give those who are struggling a chance to reason the questions out.

## 3. Discussion of *Taking Some Out, Part I*

In discussing Question 2, emphasize that the two expressions for the situation, probably 45 – (8 + 11) and 45 – 8 – 11, must be equal.

For Question 3, students will probably come up with $60 - (9 + X)$ for the rule with parentheses and $60 - 9 - X$ (or $51 - X$) as the rule without parentheses. As before, emphasize that the two expressions must be equal.

### • *Generalizing the problems*

Ask the class if they can come up with a generalization of these problems that has no numbers in it. If needed, suggest that they replace the number of cubes the first chef removed by one variable, the number the other chef removed by a second variable, and the initial amount put in the cauldron by a third variable. The goal is to obtain a general statement something like this one.

$$A - (B + C) = A - B - C$$

Also have students try to put the idea behind this formula into words. For example, a student might describe this by saying, "Subtracting a sum is like subtracting each part of the sum." Suggest to students that they can refer to this process as "distributing the subtraction." You might have them discuss its similarity to the distributive property.

## *Homework 18: Subtracting Some Sums*

This assignment reinforces the mechanics of *Taking Some Out, Part I,* and introduces some initial work on solving equations involving subtraction and parentheses. Students will probably use intuitive methods in working with the equations. More formal manipulations of equations will be introduced beginning on Day 20.

The assignment also continues to strengthen students' ability to work with the distributive property and to solve simple equations.

# *Taking Some Out, Part I*

Do you remember the chefs from *Patterns* in Year 1? You used their situation to help with the arithmetic of positive and negative numbers. Well, thinking about temperatures can also be of help when finding equivalent expressions.

In each of the problems here, you should assume that when the action begins, the temperature of the cauldron is 0 degrees. As usual, every hot cube added to the cauldron increases the temperature by one degree and every hot cube removed from the cauldron lowers the temperature by one degree.

1. The chefs decided to put 50 hot cubes into the cauldron, but once they did so, they found that the cauldron was too hot. So two of the chefs reached in and removed some hot cubes. One chef removed a batch of 5 hot cubes and the other chef removed a batch of 10 hot cubes.

   a. What was the temperature when this was all done?

*Continued on next page*

    b.  Write the entire process as a chef instruction in two ways:

- With parentheses, showing the two batches of cubes being removed together

- Without parentheses, showing the two batches of cubes being removed one batch at a time

2. Another time the chefs put 45 hot cubes into the cauldron and again found that the cauldron was too hot. Two chefs removed some hot cubes. One chef took out a batch of 8 hot cubes and the other took out a batch of 11 hot cubes.

    a.  What was the temperature at the end of this process?

    b.  As in Question 1b, write the entire process as a chef instruction in two ways.

3. The next time this happened, the chefs put in 60 hot cubes to begin with, and again two chefs took some out. The first chef removed a batch of 9 hot cubes, but the second chef forgot to count the number of cubes he removed.

    a.  Create several rows for an In-Out table in which the *In* is the number of hot cubes the second chef removed and the *Out* is the final temperature.

    b.  Find two rules for the table—one with parentheses and one without parentheses. Use $X$ to represent the *In*.

    c.  Graph your two rules on the graphing calculator and see if they give you the same graph.

# Homework 18      Subtracting Some Sums

1. Write each of these expressions as an equivalent expression without parentheses. Simplify your results where you can by combining like terms.

   a. $35 - (3a + 14)$

   b. $50 - (c + 17 + 2d)$

   c. $16 + 9s - (3s + 11)$

   d. $23 + 5w - 2(w + 7)$

2. The equations in the next series involve subtracting a sum that is in parentheses. Use whatever techniques make sense to you to solve these equations, but write an explanation of what you do.

   a. $54 - (t + 5) = 32$

   b. $29 - 2(x + 4) = 5$

   c. $6z + 17 - (2z + 5) = 56$

*Continued on next page*

3. Because the distributive property is so important in working with algebraic expressions, it's helpful to be able to apply it smoothly and with confidence. Find each of the products below, writing the results without parentheses and combining like terms when possible.

   a. $(x + 4)(x + 7)$

   b. $(2t + 3)(3t - 5)$

   c. $(4r - 3)(3r - 2)$

   d. $(x^2 + 3x + 2)(x + 6)$

4. Soon you'll be combining ideas about equivalent expressions with your insights learned from work with the mystery bags in order to solve more complex equations. In preparation for that, here are some mystery bag problems for you. Solve these equations to find out how much gold is in each mystery bag.

   a. $26M + 37 = 19M + 58$

   b. $46a + 95 = 83a + 29$

   c. $153x + 149 = 327x + 73$

# DAY 19
# *Taking Some More Out*

*Students continue work with simplifying expressions and removing parentheses.*

## Mathematical Topics

- Continuing work with equivalent expressions
- Subtracting expressions in parentheses

## Outline of the Day

### In Class

1. Discuss *Homework 18: Subtracting Some Sums*
   - Look for a variety of approaches to both removing parentheses and solving equations

2. *Taking Some Out, Part II*
   - Students work with the hot-and-cold-cube model to understand subtraction of a difference
   - The activity will be discussed on Day 20

### At Home

Homework 19: Randy, Sandy, and Dandy Return

## 1. Discussion of *Homework 18: Subtracting Some Sums*

On Question 1, you can start by letting students compare results in their groups. If they seem to be comfortable with these problems and agree on the answers, you might skip Questions 1a and 1b. But have students present Questions 1c and 1d, because these problems add new elements to the work from yesterday, both in combining nonnumerical terms (in Questions 1c and 1d) and working with a factor outside the parentheses (in Question 1d). Try to get students to build on yesterday's metaphor, applying the idea that subtracting a sum involves subtracting each part of the sum.

*"Can you simplify the expression?"*

*"Why is your simplified expression equivalent to the original?"*

On Question 1c, students will probably have a first step that looks like $16 + 9s - 3s - 11$. But they should simplify this expression, combining terms to get $5 + 6s$. Although students have combined terms before using the mystery bags metaphor, you might want to point out that writing $9s - 3s$ as $6s$ can also be seen as an application of the distributive property. You may also need to discuss order-of-operations issues to explain, for example, why students can't write $16 + 9s$ as $25s$.

On Question 1d, some students might initially write this expression as $23 + 5w - (2w + 14)$ while others might go straight to the step $23 + 5w - 2w - 14$. There are other possibilities as well, and you should encourage variety. Again, be sure that students simplify their answers.

## • *Question 2*

*"How did you come up with the number 17 in the first place?"*

On Question 2, have students focus on the reasoning they used to solve the equations. If someone says, for example, that the answer to Question 2a is $t = 17$, because $54 - (17 + 5) = 32$, ask, "How did you come up with the number 17 in the first place?"

Some students may first simplify the expression on the left and then solve. For example, on Question 2b, they might first restate the equation as $29 - 2x - 8 = 5$, next simplify to $21 - 2x = 5$, and then proceed from there.

Others may say something like, "If you subtracted from 29 and got 5, you must have subtracted 24, so $2(x + 4)$ must be equal to 24," and then solve the equation $2(x + 4) = 24$ to get the final answer.

As with Question 1, encourage a variety of approaches, trying to promote the idea that whatever students do should make sense to them.

## • *Question 3*

Probably the main issue in Question 3 will have to do with signs. You might suggest that students think of subtraction in terms of adding the opposite. For example, in Question 3b, they can think of $3t - 5$ as $3t + (-5)$, so that the partial products are $2t \cdot 3t$, $2t(-5)$, $3 \cdot 3t$, and $3(-5)$. You may also have to help students see that $2t \cdot 3t$ should be simplified as $6t^2$.

In Question 3d, you might want to remind students of the area model as a way of seeing that there are six partial products in this problem.

## • *Question 4*

The purpose of using large numbers in these problems is to have students think about these equations in terms of manipulations rather than use trial and error. Encourage them, however, to continue to think in terms of the pan-balance model.

## 2. *Taking Some Out, Part II*

This activity continues with yesterday's metaphor but moves on to the more difficult situation in which the expression in parentheses involves subtraction.

As in *Taking Some Out, Part I,* it's probably advisable to do Question 1 as a whole class to clarify the intended process.

*As a hint to get 12 − 4: "How many cubes were removed altogether?" and "Where does that number 8 come from?"*

The key idea is that the first expression in Question 1b should be 75 − (12 − 4). You might get the subtracted expression, 12 − 4, by asking how many cubes were removed altogether and then asking where the number 8 comes from. The second expression for Question 1b should be 75 − 12 + 4, showing the initial 75 cubes put in, the removal of 12 cubes, and then the return of 4 of them.

As before, the important idea is that the two expressions, 75 − (12 − 4) and 75 − 12 + 4, must be equal.

With this introduction, have students get to work on the activity, which will be discussed tomorrow.

Question 2 is quite similar to Question 1. On Question 3, students might use any of these expressions:

- 54 − (25 − 6 − 4)
- 54 − [25 − (6 + 4)]
- 54 − 25 + 6 + 4
- 54 − 25 + (6 + 4)

If you see something like 54 − 25 + 10, you can ask students to write this in more detail so that each part of the problem is visible.

## *Homework 19: Randy, Sandy, and Dandy Return*

Tonight's assignment brings back our friends from *Homework 13: Why Are They Equivalent?* and continues the work with both the distributive property and equation solving.

# Taking Some Out, Part II

The chefs are continuing to play around with their cauldron. Use each of these problems to investigate ways to write different expressions for the same situation. As in *Taking Some Out, Part I,* each situation begins with a temperature of 0 degrees.

1. The chefs tossed 75 hot cubes into the cauldron. A few minutes later, one of the chefs reached in to remove some of them. She already had 12 hot cubes in her hands when she stumbled and 4 of those hot cubes fell back in.

   a. What was the temperature at the end of this process?

   b. Using the numbers in the problem, write the entire process in two ways.

   • Show the initial amount put in and subtract an expression in parentheses to show what was removed altogether.

   • Show the initial amount put in, use subtraction to show the whole batch of cubes being removed, and use addition to show some of that batch going back in. Your expression should not have parentheses.

*Continued on next page*

2. The next time, 62 hot cubes were put in originally. A chef then removed 14 of them, but 9 of the 14 fell back into the cauldron.

    a. What was the temperature at the end of this process?

    b. As in Question 1b, write the entire process in two ways.

3. The third time, 54 hot cubes initially were tossed in. A chef reached in and grabbed 25 of them, but first 6 of the 25 and then 4 more fell back in.

    a. What was the temperature at the end of this process?

    b. Write this entire process in *at least three* different ways.

4. Once again, the chefs tossed a big batch of hot cubes into the cauldron. Someone reached into the cauldron and pulled a handful of them out, but part of that handful fell back in. Write two different general expressions, one with parentheses and one without, describing what happened. Use different variables to represent the initial amount put in, the amount initially removed, and the amount that fell back in.

# Homework 19

# Randy, Sandy, and Dandy Return

## *Part I: Generalizing the Distributive Property*

Randy, Sandy, and Dandy are having another of their heated arguments. This time they aren't discussing why the distributive property is true, but are trying to find other principles that might be based on similar reasoning.

1. Randy says, "I use the distributive property all the time, even when it just involves multiplication." In other words, Randy thinks that $a(bc) = (ab) \cdot (ac)$.

   Is she correct? Try substituting some numbers to find out. If she's right, explain why. If she's wrong, rewrite the right side of the equation to make her statement correct.

2. Sandy then says, "I use the distributive property all the time, too." (This makes Randy worry a little about what she thinks.) "And I use it with additions all over the place." What Sandy thinks is that $a + (b + c) = (a + b) + (a + c)$.

   Is she correct? Try substituting some numbers to find out. If she's right, explain why. If she's wrong, rewrite the right side of the equation to make her statement correct.

3. Dandy thinks they are both confused and says, "I don't know what you two are thinking of, but I know that $a - (b - c) = a - b + c$."

   Is he correct? Try substituting some numbers to find out. If he's right, explain why. If he's wrong, rewrite the right side of the equation to make his statement correct.

## *Part II: Distributing Mystery Bags*

4. For each of the equations shown here, first use the distributive property (correctly!) to remove the parentheses on each side of the equation, and then combine terms on each side and solve the resulting equation.

   a. $4(M + 2) + 7 = 6(M + 1) + 2$

   b. $3(x + 9) + 2(3x + 4) = 7(x + 11)$

   c. $5(2x - 3) + 3(x + 8) = 4(x + 6) + 3(x - 4)$

---

60                                                  Interactive Mathematics Program

# Equivalent Equations

*Students begin formal work with equivalent equations.*

## Mathematical Topics

- Evaluating variations of the distributive property
- Subtracting expressions in parentheses
- Using equivalent equations

## Outline of the Day

### In Class

1. Discuss Part I of *Homework 19: Randy, Sandy, and Dandy Return*

2. Discuss *Taking Some Out, Part II* (from Day 19)

   - Get students to state a general principle for subtracting a difference

3. Discuss Part II of *Homework 19: Randy, Sandy, and Dandy Return*

4. Introduce the concept of equivalent equations

### At Home

*Homework 20: Equation Time*

## 1. Discussion of Part I of *Homework 19: Randy, Sandy, and Dandy Return*

We suggest that you discuss only Part I of the homework now and follow that up with discussion of yesterday's activity, *Taking Some Out, Part II*.

You can then discuss of Part II of last night's homework, which will lead in smoothly to today's topic, equivalent equations.

You can have heart card members of different groups present their work on the questions in Part I, although you may want to include discussion of Question 3 as part of the discussion of *Taking Some Out, Part II.*

Finding counterexamples for Randy's and Sandy's ideas should not be difficult. Be sure to ask how students might modify their equations to get a correct general principle. Here are the "standard" ways to complete these equations correctly:

- Question 1: $a(bc) = (ab)c$
- Question 2: $a + (b + c) = (a + b) + c$

Tell students that these two principles are called the **associative properties** for multiplication and for addition.

Other answers are possible as well. For instance, a student might complete Randy's equation by writing $a(bc) = (bc)a$.

## 2. Discussion of *Taking Some Out, Part II*

*"What does (14 − 9) represent in the problem?"*

*"Why do you add 9 in the expression 62 − 14 + 9?"*

Have volunteers present their ideas about Questions 2 and 3 of *Taking Some Out, Part II* (assuming that you did Question 1 as a whole class yesterday). The presentation on Question 2 will probably follow the model of Question 1. Thus, students will probably get 62 - (14 - 9) and 62 - 14 + 9 as the two expressions. Insist that they describe what these expressions represent.

As noted preceding the activity, students might use several possible expressions for Question 3. Get as many different representations as students can provide.

Question 4 of *Taking Some Out, Part II,* and Dandy's suggestion in last night's homework involve essentially the same issue. You might try to get students to describe this phenomenon in a brief, intuitive way, such as, "If you're taking less out, then it's like putting some back in." Students should see that Dandy's general principle is correct and that either side of his equation could be used to describe the situation in Question 4 of *Taking Some Out, Part II.*

### • Some other possible "distributive properties"

Before leaving the issue of the validity of principles like the distributive property, point out to students that they have been working with addition, subtraction, and multiplication, but not division.

*"Are these equations true for all numbers a, b, and c?"*

In that context, put forth these two general principles as candidates:

$$(a + b) \div c = (a \div c) + (b \div c)$$

and

$$a \div (b + c) = (a \div b) + (a \div c)$$

If students aren't sure what to do, suggest that they try some numerical examples and then think about what's going on. Be sure they recognize that the first of these is correct but that the second is not.

*"Are these equations true for all numbers a, b, and c?"*

You might pose the same questions writing the expressions using fractions instead of division. That is, ask whether these principles hold:

$$\frac{a + b}{c} = \frac{a}{c} + \frac{b}{c}$$

and

$$\frac{a}{b + c} = \frac{a}{b} + \frac{a}{c}$$

It should be interesting to see if students realize that this pair of equations is essentially the same as the pair with the division signs and if they have a different sense of whether the equations are true when they are written in fraction form.

## • Practice with manipulations

You may want to give students one or two practice problems each day for the rest of the unit. This will enable them to move gradually toward a greater comfort level and increased understanding of manipulations involving removing parentheses, combining terms, and simplifying expressions.

### • Optional: Subtraction using multiplication by –1

Another approach to subtracting expressions with parentheses is to treat subtraction as addition with a factor of –1. For example, the expression $a - (b - c)$ can be thought of as

$$a + (-1)[(b + (-1)c]$$

This transition can build on the students' understanding from the hot-and-cold-cube model that subtracting a number gives the same result as adding its opposite. The general principle here is this:

$$x - y = x + (-1)y$$

Students may understand this principle best in terms of the hot-and-cold-cube model: Taking out $y$ cubes of any kind is like putting in $y$ cubes of the opposite kind. Then, using the distributive property, the expression $a + (-1)[b + (-1)c]$ can be rewritten as $a + (-1)b + (-1)^2c$, which simplifies to $a - b + c$.

Some students may find this more formal approach more appealing than the metaphor of the chefs and the cauldron.

## 3. Discussion of Part II of *Homework 19: Randy, Sandy, and Dandy Return*

Let diamond card students present each of the questions from Part II of last night's homework. Students should see that these are essentially nothing more than very complicated mystery bags problems. Although the use of subtraction in Question 4c doesn't exactly fit the metaphor, at this point students will probably be sufficiently comfortable with the mechanics of symbol manipulation that they will not be troubled by this.

We suggest that you use Question 4c to introduce the concept of equivalent equations. You can let a student present the problem, but save the work so you can refer to it in the discussion, as described in the next section.

## 4. Equivalent Equations

The concept of equivalent equations is the theoretical basis for all the algebraic techniques for solving equations. The terminology described here provides a framework for discussing these techniques.

You can use the discussion of Part II of last night's homework as a lead-in to the concept of equivalent equations. But before getting into the more general concept of equivalent equations, first review the definition of *equivalent expressions*.

For example, suppose students began Question 4c by rewriting the equation as

$$10x - 15 + 3x + 24 = 4x + 24 + 3x - 12$$

Ask students why they thought it was okay to write the equation this way. They will probably say something like, "It's the same as the other equation, just written differently."

*"Who can restate what 'equivalent expressions' means?"*

Ask what the formal term is for writing the same thing a different way. Students should recall the phrase *equivalent expressions* and be able to explain what it means—that is, these are expressions that give the same numerical result no matter what numbers are substituted for the variables.

Use the concept of equivalent expressions to continue working through the equation in Question 4c. This should bring you to an equation like

$$13x + 9 = 7x + 12$$

Ask for a volunteer to explain one step of what he or she did next, in terms of the mystery bags model. For example, suppose the student took 9 ounces of weight off both sides of the pan balance, giving the equation

$$13x = 7x + 3$$

*"Why is it correct to change the equation this way?"*

Again, ask why it's okay to make this change in the equation. Emphasize that the goal is to solve the original equation, not some other equation. This should elicit a response like, "This equation has the same solution as the original one." Tell students that equations that have the same solution are called **equivalent equations.**

Point out that many equations, such as those in the mystery bags problems, have only one solution. If two equations each have only one solution, then being equivalent simply means having the same solution.

Remind students as well that some equations have more than one solution, and bring out that this complicates the idea of equivalence. A pair of equations like $x^2 = 9$ and $x = 3$ will help clarify that having a solution in common does not make two equations equivalent. Inform students that because of this complication, we usually formulate the definition something like this:

> **Two equations are called *equivalent* if every solution of either one is also a solution of the other.**

## • Ways to get equivalent equations

*"What did you do with the mystery bags problems to get equivalent equations?"*

*"How can you state this using subtraction as a principle for getting equivalent equations?"*

Ask students what they learned from the mystery bags problems about ways to get equivalent equations. They will probably focus on the idea of taking the same thing off both sides of the pan balance. Try to get them to restate this in terms of subtraction as a principle for getting equivalent equations. For example, they might say, "Subtracting the same thing from both sides of the equation gives an equivalent equation."

Remind students of the last step in the mystery bags problems, which usually involves moving from an equation like $5M = 20$ to the solution, $M = 4$. Ask how this step might be expressed in terms of an equation. Help them as needed to state this as "Dividing both sides of the equation by the same thing gives an equivalent equation."

Then ask students what other things they can do with an equation to get an equivalent one. You might suggest equations such as $x - 4 = 7$ or $\frac{1}{2}x = 9$ to lead them to similar statements for the other operations.

It is very important that the use of these principles be intuitively clear to the students. They should feel that these rules are theirs.

You should end up with a list like this one of things one can do to get an equivalent equation.

- Replace an expression with an equivalent expression
- Add the same thing to both sides of an equation
- Subtract the same thing from both sides of an equation
- Multiply both sides of an equation by the same number
- Divide both sides of an equation by the same number

(See the next subsection for discussion of multiplying or dividing by zero.)

You may find it helpful for future discussions to come up with short-hand names for each of the principles just listed. These need not be standard terminology—let the class have fun giving its own unofficial labels to the concepts.

Students also may suggest some incorrect principles, such as squaring both sides of an equation. If so, you should give (or ask for) examples to show why the procedure does not necessarily give an equivalent equation. For example, if a student mentions squaring both sides, ask what happens if both sides of the equation $x = 3$ are squared. Bring out that the resulting equation, $x^2 = 9$, has *two* solutions while the original equation only has one. Thus, the two equations are not equivalent.

- *Watch out for zero*

Bring out that for the principles involving multiplying and dividing (the last two items in the list), the number involved must be nonzero. Students already know that they can't divide by zero, so the case of division shouldn't cause any confusion.

You may want to give an example for multiplication. For example, ask what the solution is for the equation $x + 1 = 5$, and then ask what the result is if this equation is multiplied by 0. Students should see that this simply gives $0 = 0$. They may not be sure what this means, but they should at least recognize that this equation is not equivalent to $x + 1 = 5$.

Students may not realize that they can't arbitrarily multiply or divide by expressions with variables, but that distinction shouldn't be pursued unless it comes up.

## Homework 20: Equation Time

Tonight's assignment gives students some opportunity to apply the concept and techniques of equivalent equations.

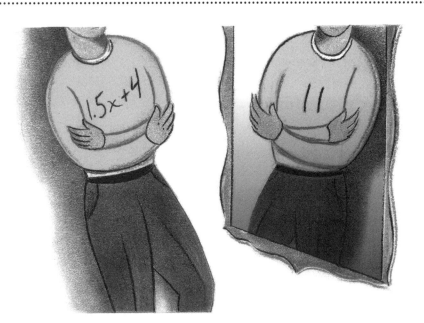

# Homework 20          Equation Time

Equations play an important role in mathematics, and the concept of equivalent equations is a valuable tool in solving equations. These problems give you a chance to apply some of what you know about equivalent equations.

1.  A student trying to solve the equation $1.5x + 4 = 11$ wrote what is shown below. Is this correct? If not, why not?

    $$1.5x + 4 = 11$$

    $$3x + 4 = 22 \quad \text{(multiplying both sides by 2)}$$

    $$3x = 18 \quad \text{(subtracting 4 from both sides)}$$

    $$x = 6 \quad \text{(dividing both sides by 3)}$$

2.  In your work with similar triangles, you have used a particular type of equation called a *proportion*, which is a statement that two ratios are equal.

*Continued on next page*

For example, if you know that the two quadrilaterals shown here are similar, you might come up with the proportion

$$\frac{x}{5} = \frac{x+2}{8}$$

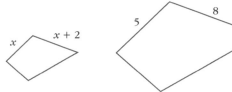

One of the principles for getting equivalent equations is that you can multiply both sides of the equation by the same thing.

a. Apply this principle to the equation $\frac{x}{5} = \frac{x+2}{8}$, first multiplying both sides by 5 and then multiplying both sides of the result by 8 (or simply multiply both sides by 40).

b. Simplify and solve the equation you got in Question 2a and check your solution in the original equation.

3. Solve each of these equations, explaining each step clearly.

a. $41 + 7d - 5(d + 7) = 8d + 1$

b. $8w - 3(2w - 9) = 7(w + 2)$

# Scrambling Equations

Students continue work with equivalent equations, going backwards from simple equations to complicated ones.

## Mathematical Topics

- Recognizing a common error in working with equivalent equations
- Working with equations involving fractions
- Continuing work with equivalent equations

## Outline of the Day

### In Class

1. Form new random groups
2. Discuss *Homework 20: Equation Time*
   - Use Question 2 to discuss the idea of cross multiplying
3. *Scrambling Equations*
   - Students turn simple equations into complex equivalent equations and then try to retrace one another's steps
   - No whole-class discussion of this activity is needed

### At Home

*Homework 21: More Scrambled Equations and Mystery Bags*

## 1. Forming New Groups

This is an excellent time to place the students in new random groups. Follow the procedure described in the IMP *Teaching Handbook,* and record the groups and the suit for each student.

## 2. Discussion of *Homework 20: Equation Time*

You can have club card students present each of the problems. Question 1 represents an opportunity to emphasize that multiplying one side of an equation by something means multiplying the *entire expression* by that factor. You might suggest to students that they insert parentheses for clarity, writing the step of multiplying both sides by 2 as $2(1.5x + 4) = 2(11)$. This will help them see that the left side becomes $3x + 8$ and *not* $3x + 4$.

This is a *very common* type of mistake, often seen even in college calculus classes, and so you should go over Question 1 carefully. Perhaps have several students articulate the principle involved here in different ways.

You can use students' work with the mystery bags to help clarify this. If the Jester wanted to make one side of the pan balance twice as heavy, then he would have to make *the entire other side* twice as heavy as well in order to keep the pans in balance. This would mean doubling both the number of mystery bags and the amount of lead weight.

Question 2 provides a context for talking about cross multiplying in a meaningful way. Although this may have been discussed briefly in the Year 1 unit *Shadows,* most students will benefit from seeing it again in the broader context of equivalent equations.

The discussion of Question 3 will give a sense of how well students can handle the mechanics developed in *Taking Some Out, Parts I* and *II.*

## 3. *Scrambling Equations*

Today's activity, *Scrambling Equations,* is designed to show equation solving as an "uncomplicating" process and to reinforce the notion of equivalent equations. Each student will take a simple equation, such as $x = 5$, and write a series of more complex equations in which each is equivalent to the preceding one.

When all students are ready, you might have them pair up within their groups to check that their work is correct. Then each example should be written on a separate sheet of paper, with the original equation (such as $x = 5$) written on one side of the paper and the final, "scrambled" equation written on the reverse side.

Each group will then trade its set of examples with another group, with the sides showing the final equations face up. Either as a whole or in pairs, groups can then work backwards to retrace the steps for each of the other group's scrambled equations until they come up with the same original equations.

It is very important for students to follow some guidelines for scrambling equations or they will go off on some peculiar tangents that no one, even themselves, could trace backwards.

## • *Introducing the activity*

Go over the rules of the activity, pointing out that there are four methods that students can use to scramble their original equations.

- They can add the same integer to both sides of the equal sign.
- They can subtract the same integer from both sides of the equal sign.
- They can multiply both sides of the equal sign by a nonzero integer.
- They can divide both sides of the equal sign by a nonzero integer.

Point out to students that they are also allowed to do numerical simplifications to the right side of the equation, but other than that, they are to stick to the four basic methods. Tell students as well that they are to scramble their original equations using any *three* of these four steps. Finally, emphasize that they can use only integers as the numbers they add or subtract or by which they multiply or divide.

The guidelines can be best explained by actually scrambling an equation for the class. Suppose you start with the equation $x = 5$.

Ask students what they want to do first. If they say, "Subtract 2," write

$$x - 2 = 5 - 2$$

and have them simplify the right side to get

$$x - 2 = 3$$

Again, ask them what they want to do next. If they say, "Multiply by 8," write

$$8(x - 2) = 8 \cdot 3$$

and have them simplify the right side to get

$$8(x - 2) = 24$$

Emphasize that they should leave the expression $8(x - 2)$ as it is and *not* multiply through to get $8x - 16$.

Suppose they suggest as their third and final step to divide by 6. Remind them that the side with the $x$ should show the steps, so their next equation will be

$$\frac{8(x - 2)}{6} = \frac{24}{6}$$

and the right side will simplify to give the equation

$$\frac{8(x - 2)}{6} = 4$$

Have students check that the original value for $x$ (namely, $x = 5$) fits this final equation.

Tell them that all they will get from another group will be a set of final equations, each like this one, and that they should try to come up with the original equations and the steps by which the scrambled equations were created.

You might show an example of this as well. For instance, give them the equation

$$\frac{2(x + 7)}{3} = 5$$

and ask them to work backwards to reconstruct the original equation. They should be able to go from this scrambled equation to $2x + 7 = 15$ to $2x = 8$ and finally to $x = 4$.

Warn students to be kind to one another! The idea is to give one another practice with this uncomplicating process, not to send each other's minds spinning.

## • *Have groups begin*

Have students begin scrambling their own equations. Once the equation is messed up, they should put the last step on a clean sheet of paper with the answer on the back.

When two groups are ready, they can exchange equations. Students solving equations should do their work on a different sheet of paper from the one they are given, so that more than one person can work the problem. Some students will make mistakes when creating their equations, so it is important that they check their answers using substitution.

Remind students that they should try to reconstruct the creator's sequence of equations, or at least get back to where they can solve the equation.

You probably don't need any whole-class discussion of this activity. However, if there are cases in which a group can't reconstruct the original equation, you can have the group put the scrambled equation on the board for everyone to work on.

## *Homework 21: More Scrambled Equations and Mystery Bags*

Part I of tonight's homework is a continuation of the work done in class today. Students should be able to work successfully on it even if they only had a little bit of time on the activity. Part II continues the work with mystery bags equations.

# Scrambling Equations

Usually, the concept of equivalent equations is used to make things simpler. But in this activity, you're going to make things more complicated. For example, look at the sequence of equations shown below.

$$x = 1$$

$$6x = 6$$

$$6x - 3 = 3$$

$$\frac{6x - 3}{2} = 1.5$$

All of these equations are equivalent, because they all have the same solution. You should be able to see what was done to each equation to get the one below it.

In this activity, you will begin by writing down a *very simple* equation (like $x = 1$). Then you'll write down an equivalent equation that's more complicated, and then something equivalent to that, and so on.

*Continued on next page*

This activity has some very precise rules. You will be changing your equation exactly three times. At each stage, you can do any one of these four things.

- You can add the same integer to both sides of the equation.

- You can subtract the same integer from both sides of the equation.

- You can multiply both sides of the equation by the same nonzero integer.

- You can divide both sides of the equation by the same nonzero integer.

Remember that you are to do *exactly three* of these steps (in any order). For instance, the example shown above uses multiplication, then subtraction, and then division. At any time in the process, you're also allowed to do arithmetic steps to simplify the right side of the equation.

When you are done with this process, copy your final, complicated equation onto one side of a sheet of paper and put your original equation on the reverse side. This sheet will be exchanged with another group, and you will have the opportunity to "uncomplicate" someone else's scrambled equation.

# Homework 21   More Scrambled Equations and Mystery Bags

## *Part I: More Scrambled Equations*

This assignment involves the same steps for getting equivalent equations that were described in *Scrambling Equations*.

1.  The equations here show one sequence of three steps to "scramble" the equation $x = 3$.

$$x = 3$$

$$x - 5 = 2$$

$$10(x - 5) = -20$$

$$\frac{10(x - 5)}{4} = -5$$

   a.  Describe what was done at each step.

   b.  Check that $x = 3$ is a solution to the final equation in the sequence, and show your work.

*Continued on next page*

Interactive Mathematics Program                                                        65

For Questions 2 through 4, do two things

- "Uncomplicate" each equation until you get back to a simple equation of the form "$x$ = some number."

- Take the value of $x$ you get from the simple equation and substitute it back into the original equation in order to check that it makes the "complicated" equation true.

2. $3x - 5 = -2$

3. $\frac{x - 6}{4} + 1 = 7$

4. $4\left(\frac{x}{3} + 6\right) - 8 = 20$

# Part II: More Mystery Bags

Earlier in this unit, you used the idea of a pan balance to solve mystery-bag problems.

The equations here might come from such problems. Solve them using the concept of equivalent equations, but also think about how each step you do is related to the pan-balance model.

5. $11t + 13 = 7t + 41$

6. $12 + 7w = 4w + 21$

7. $8(x + 3) + 19 = 15 + 2(x + 35)$

*Solve It!*

# *The Linear World*

*This page in the student book introduces Days 22 through 26.*

A straight line is one of the simplest types of graphs. In *Homework 11: Line It Up,* you saw that certain algebraic expressions lead to graphs that are straight lines.

Linear equations and linear functions are important in many applications of mathematics, and you already know a great deal about them. In an activity called *Get It Straight,* you'll work with your classmates to learn even more.

***Terria Galvez and Mekea Harvey continue their work on linear functions and straight-line graphs.***

Interactive Mathematics Program                                                                                    67

# *Summarizing Linear Equations*

*Students see that every linear equation in one variable can be solved using equivalent equations.*

## Mathematical Topics

- Summarizing methods for solving linear equations
- Defining the category of linear equations
- Seeing that algebraic methods will solve any linear equation

# Outline of the Day

## In Class

1. Select presenters for tomorrow's discussion of *POW 3: Divisor Counting*
2. Discuss *Homework 21: More Scrambled Equations and Mystery Bags*
3. Introduce the concept of linear equations
   - Bring out that the techniques students already know can be used to solve all linear equations in one variable

4. *Old Friends and New Friends*
   - Students solve familiar equations using algebraic techniques
   - The activity will be discussed on Day 23

## At Home

*Homework 22: New Friends Visit Your Home*

## 1. POW Presentation Preparation

Presentations of *POW 3: Divisor Counting* are scheduled for tomorrow. Choose three students to make POW presentations, and give them overhead transparencies and pens to take home for preparing presentations. You may want to suggest that they begin with their results from *Homework 17: Exactly Three or Four,* because these special cases probably weren't fully discussed on Day 18.

## 2. Discussion of Homework 21: More Scrambled Equations and Mystery Bags

You can have spade card students present each of the scrambling problems. However, if students seemed comfortable with these ideas yesterday, you might just do the most difficult example, Question 4. Then have presentations on Questions 5 through 7 (or at least Question 7) to connect once again the techniques for equivalent equations to the balance model of the mystery bags problems.

> *Comment:* These equations are generally simpler than those in Question 4 of *Homework 19: Randy, Sandy, and Dandy Return.* The discussion here can focus more on the formal idea of equivalence.

## 3. Linear Equations

*"How would you describe the equations in these problems?"*

Ask students how they would describe the kind of expressions that appear in Questions 5 through 7 of last night's homework. They might give responses like, "You just have something times *x* and then add or subtract on each side."

As needed, review the term *linear expression* for an algebraic expression that defines a function whose graph is a straight line. (This term was introduced in the discussion of *Homework 11: Line It Up*). Students should recall that a linear expression (in some variable, say, *x*) is an expression of the form *ax* + *b,* where *a* and *b* are any two numbers.

Point out that at the time of that discussion, students had not yet worked formally with the idea of equivalent expressions. Tell them that anything *equivalent to* something of this form is also a linear expression.

*"What are some linear expressions that don't exactly look like ax + b?"*

Ask students to give you some examples of linear expressions that don't exactly look like *ax* + *b.* If necessary, give them an example or two, such as 5(*x* + 3) or 2 – 3(*x* + 7).

Introduce the term **linear equation** to mean any equation in which both sides are linear expressions in a particular variable, and bring out that the mystery bags equations were all of this type. Have students give you some more examples of linear equations; these examples can be as complex as they want. They should see that even ugly messes like

$$5(3x - 2) - \frac{4x + 1}{3} = 7x + 2 - 2(5 - 3x)$$

are considered linear equations.

*"What are some examples of nonlinear equations?"*

For emphasis, also ask students for some examples of equations that are *not* linear. These could include expressions with higher powers of the variable or expressions with the variable in a denominator or as an exponent.

### • *Ta-da! All linear equations can be solved!*

*"What is your goal when you are given a linear equation?"*

Ask students what their goal has been when they were given a linear equation. If they say, "To solve it," ask what this means. Get them to talk about finding the number that can be substituted for the variable to fit the equation.

Then ask where they stand in the effort to solve such equations. They should realize that they have learned foolproof techniques for solving any linear equation.

### • *Optional: Back to mystery bags*

You may want to illustrate this generality by returning to the mystery bags problems. You can present students with the equation $AM + B = CM + D$, perhaps calling it "the general mystery bags equation."

Explain to students that they should think of $A$, $B$, $C$, and $D$ as if they were numbers. Specifically, $A$ and $C$ represent the number of mystery bags on each side of the pan balance, and $B$ and $D$ represent the amount of weight on each side.

*"How would you express the weight of a mystery bag in terms of A, B, C, and D?"*

Tell students that their task is to find an algebraic expression in terms of these variables for the weight of a single mystery bag. If they are perplexed by the fact that they don't know which is bigger, $A$ or $C$, you can suggest that they assume for now that $A$ is greater than $C$.

They should be able to rewrite the problem in a series of steps, perhaps first as $(A - C)M + B = D$, then as $(A - C)M = D - B$, and finally as $M = \frac{D - B}{A - C}$. Once they have this solution, you can ask them to consider what would happen if $C$ were greater than $A$. Help them to see that the resulting solution, $\frac{B - D}{C - A}$, is equal to the earlier expression, $\frac{D - B}{A - C}$.

### • *Do an ugly example*

To emphasize the generality of the algebraic techniques, have the class solve an equation like the "ugly mess" given earlier. You may need to ask leading questions such as "What's a good place to start?" or "How can you simplify this part?" or "What would you do next?" The key ideas might come from just a few students, but everyone should be able to carry out the individual steps.

## 4. Old Friends and New Friends

> *Old Friends and New Friends* culminates the "solving linear equations" segment of the unit. In Part I of this activity, students look at linear equations that they have seen in the context of problems.

Students can start right in working on *Old Friends and New Friends*. Tell groups to check their answers using substitution. This activity will be discussed tomorrow. (Tonight's homework is similar to Part II of this activity, but assigning this homework before discussing the activity shouldn't present any serious difficulties.)

*Note:* The "shadow equation" in Question 4,

$$\frac{S}{S + 20} = \frac{6}{25}$$

is not in the form of a linear equation, but it is equivalent to one. Students will probably come up with a way to deal with it, based either on their experience with proportion equations in *Shadows* or on their work with the concept of equivalent equations.

## Homework 22: New Friends Visit Your Home

It is worthwhile for students to develop comfort and facility with the techniques of solving linear equations.

Tonight's homework is very traditional—it basically gives practice in solving linear equations.

*Taneesha Hardamon ponders an idea brought up by a presenter.*

Classwork *Solve It!*

# Old Friends and New Friends

## Part I: Old Friends

Over the course of this unit, you have set up equations for a variety of problems. The examples here give brief reminders of some of those problems. With each problem is an equation that might have been used to help solve the problem.

Your task in this assignment is to solve these equations. Although you could probably solve them by trial and error or by graphing, you are to solve them here using equivalent equations. Show the steps you use to get from the equations to the solutions, and check your answers by substituting them back into the original equations. *Note:* Because these problems are stated here without details, you do not need to explain how each equation fits its problem.

1. Problem: Find the payoff that would give Al a total gain of 25 points (Question 3 from *Memories of Yesteryear*).

    Equation: $25x - 75 \cdot 2 = 25$

*Continued on next page*

68 Interactive Mathematics Program

2. Problem: Find the distance around the lake (Question 2 from *Homework 4: Running on the Overland Trail*).

   Equation: $2d + 4 = 10$

3. Problem: Find the number of days the Sawyers would need to catch up (Question 3 from *Catching Up*).

   Equation: $30N = 24N + 120$

4. Problem: Find the length of Nelson's shadow (Question 2 from *Lamppost Shadows*).

   Equation: $\dfrac{S}{S + 20} = \dfrac{6}{25}$

   (*Note:* This isn't exactly in the form of a linear equation, but it is essentially equivalent to a linear equation. You might look at Question 2 of *Homework 20: Equation Time* for ideas.)

## Part II: New Friends

These equations don't come from specific problems, but that doesn't affect the algebra. Solve each of them using equivalent equations, and show the steps you use.

5. $4(t + 5) + 3 = 7t + 19$

6. $6W - (2W + 1) = 3(W - 10)$

# Homework 22

# New Friends Visit Your Home

## *Part I: Solve It!*

As you've seen, linear equations come up in many situations. Sometimes the equations are simple, and sometimes they are complicated. Use the method of equivalent equations to solve each example here, and check your solutions by substituting into the original equations.

1. $7t - 5 = 10t + 8 - 4t$
2. $6(x - 2) = 4(x + 3) - 32$
3. $8r + 1 = 12r + 27$
4. $5(w + 4) - 3(w + 2) = 3(w + 3) - (w - 5)$
5. $6g - (3g + 8) = 16$
6. $7 - 4d = 3d - 9d + 25$
7. $6 + 4(y + 2) = 10 - 4y$

## *Part II: Write It!*

Make up a situation for which the equation $5 + x = 21 - x$ might be appropriate. Identify what the variable represents in the situation you create.

# DAY 23

# POW 3 Presentations

*Students solve some linear equations and present POW 3.*

## Mathematical Topics

- Solving linear equations in one variable
- Analyzing the number of divisors of a whole number

## Outline of the Day

### In Class

1. Discuss *Homework 22: New Friends Visit Your Home*
2. Discuss *Old Friends and New Friends* (from Day 22)
3. Have presentations of *POW 3: Divisor Counting*

### At Home

*Homework 23: From One Variable to Two*

## 1. Discussion of *Homework 22: New Friends Visit Your Home*

*"What mistakes did you make while doing your homework?"*

You may want to use this assignment to assess how well students are doing with manipulating equations and symbols. As you discuss the equations, ask students for examples of mistakes they made in their homework. You may want to point out that *everybody* makes mistakes and note that this is one reason they should always check solutions. It is important for students to see what kinds of mistakes they make so that everyone can avoid them in the future.

Be sure to discuss Question 4. Students may remember that they saw a situation like this in Question 7 of *Homework 6: The Mystery Bags Game*. Bring out that although most linear equations have a unique solution, some—like this one—are true for every number while others have no solution.

For Part II, you can have several students share the problems they made up. Ask the rest of the class to verify that the situation is appropriate for the

Interactive Mathematics Program

179

equation. Students should check whether the solution to the equation really answers the question in the made-up situation.

## 2. Discussion of *Old Friends and New Friends*

> If necessary, give students a few minutes to finish up their work from yesterday. Then bring the class together to share results.

You might choose groups at random to do each of the problems, having heart card students represent their groups. Be sure that they show the step-by-step process, particularly in the more complex examples.

## 3. Presentations of *POW 3: Divisor Counting*

Have the three selected students make their presentations. As they identify any patterns or general conclusions, encourage other class members to challenge or support these conclusions.

*"Are there any numbers other than squares of primes that have exactly three divisors?"*

You can set a model for this type of challenge if students are reluctant. For example, if someone says that the square of any prime has exactly three divisors, you can ask how sure the presenter is of that fact or whether any numbers other than the squares of primes have exactly three divisors.

*"Are there any numbers other than cubes of primes that have exactly four divisors?"*

Similarly, if a student mentions that cubes of primes have exactly four divisors, you can ask if any numbers other than cubes of primes have exactly four divisors. (Numbers that are products of two different primes also have exactly four divisors.)

Students may not be able to give formal proofs of their conclusions, but you can ask them to try to explain why these statements are true.

After the presentations, have other students add any additional observations.

*Note:* One million has exactly 49 divisors and one billion has exactly 100 divisors.

> • *Some further information*
>
> Here are four conclusions that students might arrive at.
>
> • The square of any prime has exactly three divisors.
>
> • The cube of any prime has exactly four divisors.
>
> (These two conclusions generalize to the fact that the number of divisors of a power of a prime is always one more than the exponent. For example, $2^{17}$ has 18 divisors. You can ask students why they think this is so.)

- The product of two different primes has exactly four divisors.

- The square of any whole number has an odd number of divisors; a nonsquare whole number has an even number of divisors.

Formal proof of these facts requires the principle of uniqueness of factorization into primes, but students should be able to grasp the concepts intuitively. Using lots of examples will help enormously.

It is within students' reach, although difficult, to work out a formula for the number of divisors a whole number has in terms of the way it factors into primes.

## • *No new POW*

Although *Solve It!* has ten more days, there are no more POWs in the unit. Tomorrow students will begin an activity called *Get It Straight* that will last through Day 26. They will be asked to write up this investigation of linear functions in a manner similar to that used for a POW, so this activity will serve, in effect, as their final POW of the unit.

## *Homework 23: From One Variable to Two*

This assignment gives students an opportunity to summarize their work with linear equations in one variable. It will also serve as a transition to tomorrow's activity, *Get It Straight,* in which they will investigate linear functions of a single variable.

# Homework 23  From One Variable to Two

You've seen that a linear equation in one variable such as $3x + 4 = 2(x - 1)$ has a unique solution. The equation $x + 2y = 5 + 3x + y$ is also a linear equation, but it includes
two variables and has more than one solution. This assignment looks at what it means to "solve" an equation like this.

1. Begin with the simpler two-variable equation, $y = 2x + 3$.

   a. Find at least three number pairs that fit this equation.

   b. Plot the number pairs you found in part a.

2. Now use the equation $x + 2y = 5 + 3x + y$.

   a. Find at least three number pairs that fit this equation. *Hint:* Pick a number for one of the variables and then find a value for the other variable that fits the equation.

   b. Plot the number pairs you found in part a.

3. Write an equation that is equivalent to $x + 2y = 5 + 3x + y$ but that expresses $y$ in terms of $x$. In other words, your equation should have $y$ by itself on the left and an expression involving $x$ on the right. This is called **solving for $y$ in terms of $x$.** *Suggestion:* Think of the equation as a mystery-bag problem in

# *Get It Straight*

Students begin investigating the graphs of linear functions.

## Mathematical Topics

- Extending the concept of linearity to more than one variable
- Investigating linear functions of one variable and straight-line graphs

# Outline of the Day

## In Class

1. Discuss *Homework 23: From One Variable to Two*
   - Review that the term *linear* has both a geometric and an algebraic meaning
   - Extend the concepts of linear equations and linear expressions to more than one variable

2. *Get It Straight*
   - Students investigate linear functions and straight-line graphs
   - This activity will be continued on Day 25 and discussed on Day 26

## At Home

Homework 24: A Distributive Summary

---

## Discuss With Your Colleagues

Not Just $y = mx + b$

The activity *Get It Straight,* which is introduced today, leaves students lots of room to explore graphs of linear functions on their own terms. How does this compare with an approach that starts with the formal equation $y = mx + b$ and develops rules for the graphs of linear equations based on this rule? What about the fact that the activity doesn't even use the letter $m$?

---

# 1. Discussion of Homework 23: From One Variable to Two

This homework discussion will serve as an introduction to today's activity, *Get It Straight.*

*"What are some solutions to the equation y = 2x + 3?"*
*"What kind of graph does this equation have?"*

Begin the discussion by having several diamond card students offer solutions to the first equation, $y = 2x + 3$. Plot the points as they are offered on a coordinate system with equal scales for the $x$- and $y$-axes. If necessary, prod students to provide solutions in which the values of $x$ and $y$ are not both whole numbers. Then ask what kind of graph this equation has. Presumably, students will recognize the graph as a straight line.

Repeat the process with the second equation, $x + 2y = 5 + 3x + y$.

## • The meanings of "linear"

*"What does 'linear' mean?"*

Ask the class to review what *linear* means. Bring out that in *Homework 11: Line It Up,* students saw that certain types of *functions* have graphs that are straight lines and that they defined a *linear function* as one whose graph is a straight line. Emphasize that this is a *geometric* definition.

Remind students as needed that from the geometric definition, they got the concept of a *linear expression,* which is an expression equivalent to something of the form $ax + b$. Point out that this is an *algebraic* definition, but it is one that generalizes easily to more than one variable. Tell students that the expressions $x + 2y$ and $5 + 3x + y$, used in Question 2, are also considered linear expressions. Emphasize that the equation $x + 2y = 5 + 3x + y$ fits both views of what *linear* means.

## • Question 3

Finally, ask for an equation expressing $y$ in terms of $x$ for the equation $x + 2y = 5 + 3x + y$. Someone will probably come up with the equation $y = 5 + 2x$.

*Note:* Subsequent homework assignments will focus on the process of solving equations for one variable in terms of others (see *Homework 25: All by Itself* and *Homework 26: More Variable Solutions*). Therefore, you should not get sidetracked here by a general discussion of how to do this. Because the given equation is fairly simple, at least some students probably will get it. If not, you can lead them quickly to such an equation by suggesting that they subtract $x$ and $y$ from both sides.

Have students verify, for at least two or three points from their graph, that the coordinates fit this new equation as well, thus providing additional confirmation that the new equation is equivalent to the original. Point out that this equivalence brings them back to the way the term *linear* was used in *Homework 11: Line It Up* to refer to a function whose graph is a straight line. Emphasize that the linear *function* defined by the equation $y = 5 + 2x$ is just another way of looking at the linear *equation* $x + 2y = 5 + 3x + y$.

## 2. *Get It Straight*

*Get It Straight* gives students a chance to solidify their experience with straight-line graphs and to reinforce the interconnections among algebraic expressions, functions, and graphs. This is a good activity for students to do in pairs. They can select their partners and start the activity today.

The schedule of the unit gives them the rest of today as well as all of Day 25 (other than time for homework discussions) to complete their work. On Day 26, each pair will make a brief presentation of some aspect of the activity.

Tell students that in the next activity, *Get It Straight,* they will investigate how the algebra of the expression defining a linear function is related to the geometry of its graph. Then have them look over *Get It Straight.* If you think it necessary, have students brainstorm a list of various linear functions, building on the preceding discussion.

Point out to students that they will be doing a write-up similar to what they would for a POW and that the categories are two of those used in *POW 3: Divisor Counting*.

*Warning:* Some of the concepts that students may explore will be complicated by the fact that the graphing calculator does not necessarily have matching scales for $x$ and $y$. For example, although parallel lines will always appear parallel, perpendicular lines may not appear perpendicular. You can suggest to students who are investigating perpendicularity that they use windows in which a unit on the $x$-axis has the same length as a unit on the $y$-axis.

If students seem to be having a hard time getting their investigation going, you can suggest that they keep $a$ or $b$ constant and see what happens.

## Homework 24: A Distributive Summary

This assignment will become part of students' portfolios for this unit.

Classwork

# *Get It Straight*

You know that any equation involving *x* and *y* can be used to create a graph. The graph is defined as the set of all those points whose coordinates fit the equation.

Some equations have graphs that are straight lines. These are called **linear equations.** When a linear equation expresses *y* in terms of *x*, it can be referred to as a **linear function.** All linear functions can be simplified so that they fit the form *y = ax + b,* where *a* and *b* represent two numbers.

The number *a* is referred to as the **coefficient of** *x* and the number *b* is called the **constant term.** Keep in mind that these numbers can be positive, negative, or zero, and they can also be identical.

In this activity, you will investigate linear functions and straight-line graphs.

Here are some questions to explore.

- How do you change the equation in order to change the "slant" of its graph?

- How do you change the equation in order to shift the whole graph up or down?

*Continued on next page*

72

Interactive Mathematics Program

- When do two linear functions give parallel lines (lines that never meet)? Why?

- What linear functions give horizontal lines? Why?

- When do two linear functions give lines that are mirror images of each other with the *y*-axis as the mirror? Why?

- When do two linear functions give perpendicular lines (lines that form a right angle)? Why?

Do not feel limited by these questions—let your imagination soar! Keep track of any other interesting questions you think of, even if you can't answer them.

# Write-up

You should do a written report of what you learn, using these categories.

1. *Questions:* What questions did you ask yourself? Why did you ask them? Which ones did you decide to explore?

2. *Results and Conjectures:* What conjectures did you find as possible answers to your questions? What rules or patterns did you find in exploring your questions? If you can prove that your conjectures are right, do so. If you can explain why some rule or pattern works, do that as well. If possible, generalize your results.

# Homework 24 A Distributive Summary

The distributive property played an important role in this unit. This assignment on that important idea will be part of your unit portfolio.

Write as complete an explanation of the distributive property as you can. You should touch on at least these issues.

- What it says

- Why it's true

- Examples of situations in which you would use it

# DAY 25

# *Linear Functions and Straight-Line Graphs*

*Students continue investigating the graphs of linear functions.*

## Mathematical Topics

• Continuing work with linear functions and straight-line graphs

## Outline of the Day

### In Class

1. Discuss or simply collect *Homework 24: A Distributive Summary*
2. Have students continue to work on *Get It Straight* (begun on Day 24)

### At Home

*Homework 25: All by Itself*

### 1. Discussion of *Homework 24: A Distributive Summary*

You may want to have some volunteers present their ideas, or you may prefer to simply collect the assignment and allow students the full class period for work on *Get It Straight*.

### 2. Continued Work on *Get It Straight*

Have students use the rest of today's class time to work on *Get It Straight*. While they are working, begin to assign parts of the activity for pairs to present. If the class is large, you might not want to have all pairs make presentations. Give the presenters chart paper to prepare so that you can post the conclusions on the wall for awhile.

If some groups need more questions to investigate, you can suggest that they consider all possible linear equations, not just those that give $y$ in terms of $x$. For example, they might look at equations in the form $ax + by = c$ and investigate what patterns of coefficients lead to parallel lines.

## Homework 25: All by Itself

This assignment continues the idea of transforming an equation into a function by expressing one variable in terms of others. (Students did a simple example of this in *Homework 23: From One Variable to Two.*)

# Homework 25        All by Itself

You've seen that a linear equation in two variables can be transformed into a linear function by writing one of the variables in terms of the other variable. This assignment continues that theme.

1. In the activity *Fair Share on Chores* (from the Year 1 unit *The Overland Trail*), three boys and two girls were responsible for watching the animals for a total of ten hours, with the boys and the girls each having a shift of a certain length. (*Remember:* Families considered this fair in light of other chores the boys and girls had to do.)

   If $B$ represents the length of each boy's shift and $G$ represents the length of each girl's shift, then the fact that the total time is ten hours can be represented by the equation

$$3B + 2G = 10$$

*Continued on next page*

Interactive Mathematics Program      75

Solve this equation for *B* in terms of *G*. In other words, find an equivalent equation of the form

$$B = \text{an expression involving } G$$

*Hint:* One approach is to imagine that you knew the length of each girl's shift (the value of *G*) and to think about how you would figure out the length of each boy's shift (the value of *B*).

2. In Question 1, you worked with a linear equation that came from the context of a real-life situation. In these examples, all you have are the equations. In each case, solve the equation for the specified variable.

a. Solve this equation for *v* in terms of *w*.

$$2v + 7 = w - 3$$

b. Solve this equation for *s* in terms of *r*.

$$4r + 5s + 2 = 8r - s + 7$$

c. Solve this equation for *y* in terms of *x*.

$$5y - 2x + 1 = 3(y + x) - (x - 5)$$

d. Solve the equation from Question 2b for *r* in terms of *s*.

## DAY 26

# "Get It Straight" Presentations and Variable Solutions

Students present results of their investigation of the graphs of linear functions.

## Mathematical Topics

- Investigating graphs of linear functions and equations for straight-line graphs
- Solving equations for one variable in terms of other variables

## Outline of the Day

### In Class
1. Presentations of *Get It Straight* (begun on Day 24)
2. Discuss *Homework 25: All by Itself*

### At Home
Homework 26: More Variable Solutions

## 1. Presentations of *Get It Straight*

Have the student pairs assigned a part of the assignment yesterday give their presentations. Then ask other students to add conclusions they have found.

You want students to arrive at several conclusions.

- The greater the absolute value of the coefficient of *x,* the "steeper" the graph.
- A positive coefficient of *x* gives a graph that goes *up* to the right, and a negative coefficient of *x* gives a graph that goes *down* to the right.

- Linear functions with the same coefficient for *x* have graphs that are parallel lines.

- When the coefficient of *x* is zero, the graph is a horizontal line.

- There is no linear function whose graph is a vertical line. (Although the linear *equation* $x = c$ has a vertical-line graph, this equation does not represent a function.)

If students don't offer these conclusions, you should elicit them through questions. But do not worry about the formal concept of slope at this time. Students will probably have developed some intuitive insights about it as a result of their investigation, and that is sufficient for now. Also, although the idea of "mirror image" graphs is mentioned in the activity, you need not pursue that if no one brings it up. (Slope will be discussed and formally defined in the Year 3 unit, *Small World, Isn't It?*)

### • *Vertical lines*

*"What kind of linear function gives a vertical line for its graph?"*

If no one brings up the issue of vertical lines, raise it yourself, perhaps asking, "What kind of linear function gives a vertical line for its graph?" Students should see that there aren't any. Perhaps they should be able to explain as well that a linear equation expressing *y* in terms of *x* must have points with different *x*-coordinates.

> *Note:* Because of distortion or unbalanced viewing windows, some students may have obtained graphs on their graphing calculators that actually appeared vertical. Be sure students see that these graphs are really not vertical, but rather are simply very steep lines.

*"What kind of equation would give a vertical graph?"*

Ask what kind of equation—not a function—would give a vertical graph. As a hint, ask what is special about all the points on a vertical line. Help students as needed to see that these points all have the same *x*-coordinate. (If necessary, pick a particular vertical line to discuss.)

> *Note:* The supplemental problem *A Function—Not!* discusses the distinction between a function and an arbitrary set of number pairs. We suggest, however, that you wait until after the discussion of *Homework 29: Functioning in the Math World* to assign this activity.

## 2. Discussion of *Homework 25: All by Itself*

You can have club card students present their work on the various homework problems.

On Question 1, you may have some students who used the hint and others who worked more formally, using techniques of equivalent equations. You could have more than one presentation on this example. Students might write their solution in fraction form, as $B = \frac{10 - 2G}{3}$, or they might use the division sign. In the latter case, be sure that they include parentheses, writing $B = (10 - 2G) \div 3$ and *not* $B = 10 - 2G \div 3$.

Questions 2a and 2b are fairly straightforward, although students might express their answers in different forms. For example, on Question 2a, they might get $v = \frac{w - 10}{2}$ or $v = \frac{w}{2} - 5$. Be sure they realize that both are correct.

Also be alert for some common mistakes. For instance, on Question 2b, students may add and subtract terms to get $6s = 4r + 5$ and then make either of these errors.

- "Subtracting 6" from $6s$ to get just $s$, essentially treating $6s$ as if it were $6 + s$. This would yield the incorrect equation $s = 4r - 1$.

- Dividing by 6 to correctly get $s$ on the left, but then only dividing one of the terms on the right by 6 and leaving the other term unchanged. This could yield either of the incorrect equations $s = 4r + 0.83$ and $s = 0.67r + 5$.

Question 2c includes a review of subtraction of expressions in parentheses, so be alert for errors in simplification.

On Question 2d, some students may prefer to solve for the $r$ term using a positive coefficient and therefore choose to leave it on the right, rewriting the equation as $6s - 5 = 4r$. Others may insist on getting the variable for which they are solving on the left side and write the equation as $-4r = -6s + 5$.

The approaches will yield answers that appear different

$$\frac{6s - 5}{4} \quad \text{or} \quad \frac{-6s + 5}{-4}$$

and you should help students see that the answers are equivalent.

*Note:* A nonlinear equation is considered **linear in a given variable** if it would become a linear equation for that variable if the other variables were replaced by specific numbers. The supplemental problem *Linear in a Variable* extends the ideas in last night's homework to such equations.

## Homework 26: More Variable Solutions

Tonight's homework continues the topic begun in last night's assignment.

You may want to take a minute to go over the use of subscripts in Question 4.

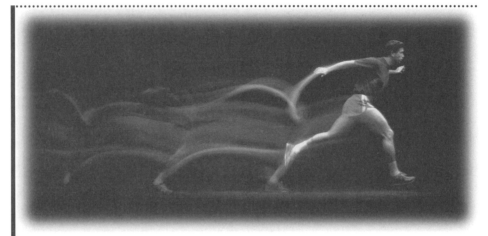

# Homework 26      More Variable Solutions

In *Homework 25: All by Itself,* you looked at some linear equations and found equivalent equations expressing one variable in terms of another. In this assignment, only Question 1 involves a linear equation.

1. Solve this linear equation for $z$ in terms of $x$.

$$3(x - z) - 2(5 - x) = 4z + 2 - 6(x + z)$$

On the remaining examples, you may find it useful to imagine that certain variables represent numbers or actually to replace them with numbers. Then think about how you would solve the given equation for the remaining variable. This approach is used in Question 2.

2. The kinetic energy of a moving object is given by the formula

$$W = \frac{1}{2} mv^2$$

where $W$ represents the kinetic energy, $m$ represents the mass of the object, and $v$ is the object's velocity.

    a. Suppose $W = 30$ and $v = 3$. What is the numerical value of $m$? (Don't worry about the units involved for energy, mass, or velocity.)

    b. Use your work from Question 2a to solve the equation for $m$ in terms of $W$ and $v$.

*Continued on next page*

3. Solve the equation from Question 2 for $v$ in terms of $W$ and $m$.

4. Coulomb's law states that the force of attraction or repulsion between two electrical charges is proportional to the product of their charges and inversely proportional to the square of the distance between them. In symbols, Coulomb's law can be expressed by the equation

$$F = \frac{kq_1q_2}{r^2}$$

where $F$ is the force, $q_1$ and $q_2$ are the charges, and $r$ is the distance between them. The letter $k$ represents a number called a **constant of proportionality.**

a. Solve this equation for $q_1$ in terms of the other variables.

b. Solve this equation for $r$ in terms of the other variables.

*Note:* The small numbers 1 and 2 in the notation $q_1$ and $q_2$ are called **subscripts.** Subscripts are often used when several variables represent similar things in a problem situation. In this case, there are two particles with electric charges, and the sizes of the charges are each represented using a single variable consisting of the letter $q$ and a subscript number.

**Days 27-31**

# Beyond Linearity

So now you can solve any linear equation! But what about equations that aren't linear? Are there any systematic ways to solve them?

In the final segment of this unit, you'll see how useful graphs and graphing calculators can be in solving more complicated equations.

***This page in the
student book
introduces Days 27
through 32.***

***Kasey Kure writes his thoughts in response to an IMP problem.***

# Where's Speedy?

*Students see how to use a formula and a graph to solve a problem.*

## Mathematical Topics

- Solving equations for one variable in terms of other variables
- Using the algebraic expression for a function to write an equation for a problem situation
- Using a graph to solve an equation

## Outline of the Day

### In Class

1. Discuss *Homework 26: More Variable Solutions*
   - Focus on the idea of equations that are not linear, but are linear in a given variable
2. Summarize work on linear equations, and make the transition to the use of graphs for more general equations
3. *Where's Speedy?*
   - Students use a formula to develop an equation for a problem situation and use a graph to solve the equation

4. Discuss *Where's Speedy?*
   - Focus on the interconnections among formulas, equations, and graphs

### At Home

Homework 27: A Mixed Bag

## 1. Discussion of *Homework 26: More Variable Solutions*

Let spade card students present their solutions to some of these problems. (You may want to keep this discussion brief in order to have time for students to complete and discuss today's activity, *Where's Speedy?*)

Question 1 does not present any new ideas, but it does provide a valuable review of the distributive property and subtraction of expressions in parentheses.

Questions 2 through 4 present an important idea for solving equations for one variable in terms of others. Students should suggest at least one example of how to use the technique of substituting numerical values for variables. For instance, on Question 2, students might replace $W$ with 4 and $v$ with 5 to get the equation $4 = \frac{1}{2}m \cdot 25$. They should be able to solve this as $m = \frac{8}{25}$ (leave it as a fraction!) and then see that the numerator is twice $W$ and the denominator is $v^2$.

Question 3 presents the first case in which a square root is needed to express the solution of an equation in terms of other variables. Although students will probably be familiar with square roots (from the Year 1 unit *The Pit and the Pendulum,* for example), you may need to review the notation of the square-root symbol.

You can point out that in terms of the equation itself, $v$ could be negative, so that there may be two solutions for $v$ for a given choice of values for $W$ and $m$. Mention as well that in the context of the physics represented by the equation, only the positive solution is needed. (A similar comment applies to Question 4b.) However, because students will work with square roots more extensively in a later Year 2 unit, *Do Bees Build It Best?* we recommend that you only briefly mention the issue of sign here.

## • *What are these formulas?*

You may want to discuss with students the issue of manipulating formulas when they don't know what the formulas mean. Most students probably will not have the physics background to understand what these equations or even the quantities in them mean, but they should not let themselves be intimidated by the unfamiliar symbols.

## 2. Beyond Linear Equations

> The final segment of the unit focuses on the use of graphing as a technique for solving equations for which algebraic manipulations may be inadequate.

As a transition to the last portion of the unit, discuss how students have been focusing on linear equations and have seen that *any* linear equation can be solved by algebraic methods using the concept of equivalent equations.

Inform students that these methods do *not* work for all equations. (You can mention that the next most complicated type of equation is the *quadratic* equation and that a Year 3 unit, *Fireworks,* will focus on quadratic equations and quadratic functions.) Tell them that graphing can yield approximate solutions to many equations that cannot be solved algebraically and that the final portion of the unit will focus on how to use graphs to solve equations.

## 3. Where's Speedy?

This assignment resumes the work on the relationship among equations, tables, and graphs, and focuses on the use of a graph to solve an equation. You may need to emphasize to students that they need not come up with a justification for the formula for Speedy's distance (and you might acknowledge that her running wouldn't fit such a formula perfectly). Students will need to change variables in order to graph the function on the graphing calculator.

No other introduction is needed.

## 4. Discussion of Where's Speedy?

Unless you saw difficulties with function notation as students worked on Questions 1 and 2, you can begin the discussion of *Where's Speedy?* with Question 3. Students might either write the equation as $0.1t^2 + 3t = 200$ or use function notation and write the equation as $m(t) = 200$.

### • Question 4

For Question 4, students should have found from their graphing calculators that the solution is approximately $t = 32$ seconds. Have one or more volunteers give details about the mechanics of this process. This can include these steps.

- Changing the variables so that the function graphed is $Y = 0.1X^2 + 3X$
- Adjusting the window settings to include $Y = 200$
- Looking for a point on the graph whose $Y$-coordinate is 200
- Finding the $X$-coordinate of this point
- Interpreting that $X$-coordinate as the desired value of $t$

*"How would you check this solution without a graph?"*

Ask students how they would check this solution without a graph. They should see that they simply need to substitute 32 for $t$ in the expression $0.1t^2 + 3t$. That is, they should find the value of $m(32)$ and check that it is approximately 200.

## Homework 27: A Mixed Bag

Tonight's assignment is a review of various ideas from the unit.

# *Where's Speedy?*

Speedy is the star runner for her country's track team. Among other things, she runs the last 400 meters of the 1600-meter relay race.

A sports analyst recently studied the film of a race in which she competed. The analyst came up with this formula to describe the distance Speedy had run at a given time in the race.

$$m(t) = 0.1t^2 + 3t$$

In this formula, $m(t)$ gives the number of meters Speedy had run after $t$ seconds of the race, with both time and distance measured from the beginning of her 400-meter segment of the race. (This formula might not be very accurate, but you are to work on this activity as if it were completely correct.)

1. Use the formula for $m(t)$ to fill in several rows of this In-Out table to show how far Speedy had run at different times of the race.

| $t$ | $m(t)$ |
|---|---|
|  |  |

2. Use the table from Question 1 to make a graph that represents this situation. You may need to add more information to your table in order to obtain a graph that shows the entire time she is running.

3. Write an equation using the variable $t$ that you could solve to get the answer to the question "How long does it take Speedy to run her first 200 meters?"

4. Graph the function $m(t) = 0.1t^2 + 3t$ on the graphing calculator and use your graph to get an approximate solution to your equation from Question 3.

# Homework 27                          A Mixed Bag

1. Find the numerical solution to each of these linear equations. Be sure to check your answers.

   a. $5a + 3 = 7a - 9$

   b. $4(t + 7) - 2(t - 3) = 8t + 11$

2. Solve each of these equations for the variable indicated in terms of the other variables.

   a. Solve for $r$ in terms of $u$.

   $$8u + 5r - 14 = 0$$

   b. Solve for $y$ in terms of $a$, $b$, $c$, and $x$.

   $$ax + by = c$$

   c. Solve for $m$ in terms of $n$.

   $$3m^2 + 5n^2 = 21$$

# DAY 28 *Graphs and Equations*

*Students continue working with graphs and equations.*

## Mathematical Topics

- Solving equations
- Continuing work with situations, functions, equations, and graphs

## Outline of the Day

### In Class

1. Discuss *Homework 27: A Mixed Bag*
2. *To the Rescue*
   - Students continue working with situations, functions, equations, and graphs
3. Discuss *To the Rescue*
   - Emphasize the use of the formula both to set up the equation and to solve the equation graphically

### At Home

Homework 28: Swinging Pendulum

## 1. Discussion of *Homework 27: A Mixed Bag*

Most of this assignment should be fairly routine to students by now, so you might focus on Questions 2b and 2c. On Question 2b, tell students that the equation is sometimes referred to as the **standard form** of a linear equation. Bring out that if *b* represents the number 0, the general solution involves division by 0. You might ask students how this connects with their work on *Get It Straight*.

You can use Question 2c to review the issue of sign, bringing out that both *m* and *n* can be negative. You might also point out that for certain values of *n*, there is no numerical value for *m* that fits the equation (and vice versa).

# 2. *To the Rescue*

*To the Rescue* continues the graphical approach to solving equations. The activity focuses on these ideas.

- Using the formula for a function to set up an equation

- Using a graph to solve an equation

- Interpreting a graphical solution in terms of both the equation and the situation

If you think students are having trouble because of the use of function notation, you can get them started with questions like "How high are the supplies after 1 second? After 2 seconds?"

# 3. Discussion of *To the Rescue*

You may want to go over Question 1 briefly to establish some background about how to use the formula for *h*.

Question 2 provides another opportunity to use a formula to develop an equation. Students should see that they simply need to set the expression for $h(t)$ equal to 100. If they write this initially as $h(t) = 100$, ask them to write it also using the formula for the function *h*—that is, as the equation $400 - 16t^2 = 100$.

On Question 3, you may need to spend some time helping students with the mechanics of using the graphing calculator to find the right value of *t* for this equation. In particular, they will probably need to change variables to enter the equation and then look for a point on the graph whose *y*-coordinate is 100.

> *Note:* The solution is $t = 4.3$ seconds, to the nearest tenth of a second.

Question 4 should be straightforward, but it is important for students to articulate the connection between the graph and the formula. They don't need to give anything more sophisticated in their answer than, "Substitute $t = 4.3$ into the expression $400 - 16t^2$ and see if you get 100." They might also say, more briefly, "Find $h(4.3)$."

- *Back to the situation*

*"What does the answer 4.3 represent?"*

Even though the equation itself was developed to answer a specific question, go back to the context and ask students to explain the meaning of their numerical solution. For example, they might say, "This means it takes about 4.3 seconds until the supplies are just 100 feet off the ground."

## • *Functions and graphs from equations*

**"How would you have solved $400 - 16t^2 = 100$ with a graph if it weren't part of the context of this problem?"**

You can reinforce the idea that any equation can be solved (approximately) with a graph by asking students what they would have done if they had been asked simply to solve the equation $400 - 16t^2 = 100$ without any discussion of the context. They may say that they would have done the same thing they just did—graph the function defined by the expression $400 - 16t^2$.

Bring out, based on their response, that they could take any expression and use it to define a function. Ask, for example, what they would graph to solve the equation $3t^2 + 5t + 7 = 10$. They should see that they could graph the function defined by the expression $3t^2 + 5t + 7$ and find the point or points on the graph where the second coordinate is 10.

Have them do this (that is, solve the equation $3t^2 + 5t + 7 = 10$ on the graphing calculator). Don't worry about how exact their answers are—the purpose here is to have students grasp the general idea of solving a somewhat arbitrary equation graphically.

## *Homework 28: Swinging Pendulum*

Tonight's homework continues the theme of the interconnections among formulas, graphs, and problem solving. As in both *Where's Speedy?* (Day 27) and today's activity, *To the Rescue,* students are given the key equation directly.

# To the Rescue

A helicopter is flying to drop a supply bundle to a group of firefighters who are behind the fire lines. At the moment when the helicopter crew makes the drop, the helicopter is hovering 400 feet above the ground.

The principles of physics that describe the behavior of falling objects state that when an object is falling freely, it goes faster and faster as it falls. In fact, these principles provide a specific formula describing the object's fall, which can be expressed this way:

> Suppose that the object's height off the ground when it begins to fall, at time $t = 0$, is $N$ feet, and use $h(t)$ to represent the object's height off the ground $t$ seconds after being dropped. Then the function $h(t)$ is given by the equation $h(t) = N - 16t^2$.

So in the case of the falling supplies, the formula is $h(t) = 400 - 16t^2$, because the supply bundle is 400 feet off the ground when it starts to fall.

1. How many seconds will it take the bundle to reach the ground? (*Hint:* What is $h(t)$ when the bundle reaches the ground?)

2. Write an equation that you could use to find out how many seconds it takes until the supply bundle is 100 feet off the ground.

3. Use a graphing calculator to find an approximate solution to your equation from Question 2.

4. Explain how you could check your solution from Question 2 using the formula $h(t) = 400 - 16t^2$.

---

82                                                                   Interactive Mathematics Program

# Homework 28                    Swinging Pendulum

In Question 4 of *Memories of Yesteryear,* you read about a group of students doing experiments with pendulums. They expressed the period of a pendulum as a function of its length by the formula $P = 0.32\sqrt{L}$, where $P$ is the time of one period (expressed in seconds) and $L$ is the length of the pendulum (expressed in inches).

1. Use the students' formula to find the period for pendulums with each of these lengths.

   a. 4 inches

   b. 16 inches

   c. 50 inches

   d. 10 feet

2. Use your answers from Question 1 and other data that you obtain from the formula to sketch a graph of the function. Use scales for your axes that are appropriate for your answers to Question 1.

After they found their formula, the students decided to build a clock using a pendulum. They wanted the period of the pendulum to be exactly one second.

3. Write an equation that they could use to find the correct length for their pendulum.

4. Find a solution to your equation. (Give an approximate value if necessary.)

5. Explain in words how you could use a graph of the function to solve the equation.

---

Interactive Mathematics Program                                                        83

# Mystery Graph

## Mathematical Topics

- Continuing work with situations, functions, equations, and graphs
- Getting information about a function from its graph

## Outline of the Day

### In Class

1. Discuss *Homework 28: Swinging Pendulum*
   - Focus on how to adapt a general formula to an equation for a specific situation

2. *Mystery Graph*
   - Students answer questions about a function using only the graph, without an equation

3. Discuss *Mystery Graph*
   - Review the term **intercept**
   - See that a graph can be thought of as an answer key to a family of equations

### At Home

*Homework 29: Functioning in the Math World*

## 1. Discussion of *Homework 28: Swinging Pendulum*

*"Why did you replace the letter P with the number 1?"*

Students can discuss the homework in their groups and then share results and difficulties. Have them check their graphs with the graphing calculators. You might point out that the pendulum in Question 1b is four times as long as that in Question 1a but that its period is only twice as long.

For Question 3, students can use an equation such as $1 = 0.32\sqrt{L}$. Although some students may begin to see questions like this as trivial, others are likely to have trouble seeing the simplicity of the question. Therefore, be sure to take some time to talk about what this equation means.

In solving the equation (Question 4), some students may realize that they can divide 1 by 0.32 and then square to find $L$. Others will use trial and error. *Note:* The correct answer, to the nearest tenth of an inch, is 9.8 inches.

Question 5 gives students another opportunity to articulate the process of using a graph to solve an equation. Try to get them to be fairly detailed in their verbal description of this process.

## 2. *Mystery Graph*

*Mystery Graph* is intended to extend students' understanding of the relationship between graphs and equations using a more abstract setting. It also provides a review and application of the function notation, $y = f(x)$.

Before letting students get started on the activity, do a simpler example to get them going. You can sketch any graph, perhaps one that is simpler than that in the activity, and ask some questions like those in Questions 1 and 3 of the activity. For example, you might sketch the graph shown here. (An enlarged copy of this graph is included in Appendix B for you to use in making a transparency.)

*Note:* Some students may recognize this as the graph of a quadratic equation, but that's not important to this discussion.

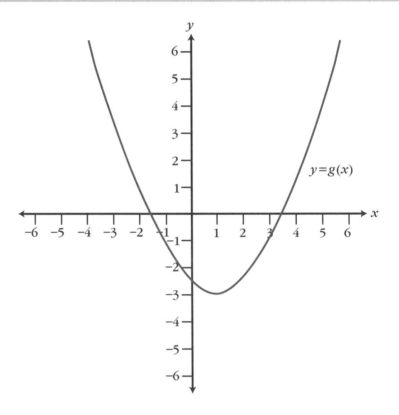

You can describe the equation $y = g(x)$ as a "generic" function equation, and ask questions such as, "What is $g(5)$?" or "What are the solutions to the equation $g(x) = -1$?" Approximate answers are fine, because your graph will probably be imprecise anyway. The goal is simply to be sure that students understand the meaning of the questions.

With this introduction, let groups begin work on the activity.

# 3. Discussion of *Mystery Graph*

Although Question 1 may seem like a straightforward exercise in graph reading, the abstract notation may still be confusing to some students. Be sure that students are clear on these basics before moving on. That is, students should see that they want to find the *y*-coordinate of the point on the graph with the given *x*-coordinate.

You can use Question 2 to bring out that students need to make some assumption about any portion of the graph that isn't visible. Students might make any one of these reasonable assumptions regarding the "missing" part of the graph.

* The graph continues upward as it leaves the visible portion.

* There is no portion of the graph that isn't visible. If this is suggested, you can use the occasion to review the term *domain* (introduced in *Patterns*). According to this assumption, the domain of the function is approximately from $x = -4$ to $x = 6$.

* The graph comes back down again or continues to fluctuate up and down.

Students who make the last of these assumptions may give different answers to Question 2 or say that there is no way to answer the question based on only partial information about the graph. You may want to agree as a class on one of these assumptions (probably the first) for the rest of the homework discussion. The discussion here makes this assumption.

## • *Intercepts*

**"What do you call the places where the graph crosses the x-axis?"**

Ask students if they know a specific term for the answers to Question 2. As needed, tell students that these values are called **x-intercepts,** because these are *x*-values for which the graph "intercepts" the *x*-axis. (*Note:* This term was introduced in passing in the Year 1 unit *The Overland Trail*.)

**"What do you suppose a y-intercept is?"**

Have students guess what the term **y-intercept** might mean and how they might express this idea using function notation. They should be able to guess that this term refers to the *y*-coordinate of the point where the graph crosses the *y*-axis and see that the *y*-intercept is the same thing as $f(0)$.

*Note:* Sometimes the term *intercept* is casually used to refer to the point itself rather than its particular coordinate. For example, one might refer to the point (5, 0) (rather than the number 5) as an *x*-intercept for the graph in this problem.

## • *Question 3: A graph is like an answer key*

On Question 3, be sure that students see that these equations have different numbers of solutions. In other words, different *y*-values can occur different numbers of times. For example, $y = 7$ occurs twice and $y = 1$ four times, while $y = -5$ does not occur at all (based on the assumption that the graph continues upward to the left beyond $x = -4$ and upward to the right beyond $x = 6$).

Use this problem as an opportunity to bring out that a graph is like an answer key to a whole family of equations—namely, those involving the expression defining the function.

## • *Questions 4 and 5*

On Question 4a, the maximum point occurs at approximately (0.5, 1.8).

On Question 4b, the minimum point appears to occur at the right end of the given interval, at approximately (3, -2.5). Students may rightly argue that the *x*-coordinate of this point isn't "between" -3 and 3; the word *between* is ambiguous. If the endpoints of the interval are excluded, there appears to be no minimum value for the function.

For Question 5, students should say something like, "Any number bigger than 5, between -1 and 2, or less than -3 will make *y* positive."

## *Homework 29: Functioning in the Math World*

This assignment summarizes some general ideas about the concept of function and continues the work of relating the different forms of representation.

*John Beguhl and David Perez compare results, one having graphed by hand, the other using the graphing calculator.*

# Mystery Graph

The graph below shows the variable $y$ as a function of $x$, but it doesn't give a formula for this function. Instead, the graph is labeled with the generic function equation, $y = f(x)$.

Answer these questions based on the graph. Give approximate answers if necessary and state any assumptions you make about any portion of the graph that isn't visible.

1. a. Find $f(4)$. That is, what number would you get for $y$ if you substituted 4 for $x$?

    b. Find $f(0)$.

    c. Find $f(-1)$.

    d. Find $f(-4)$.

2. Find all solutions to the equation $f(x) = 0$. That is, find all the values of $x$ for which $y$ is 0.

3. Solve each of these equations, giving all possible solutions.

    a. $f(x) = 7$

    b. $f(x) = 1$

    c. $f(x) = -2$

    d. $f(x) = -5$

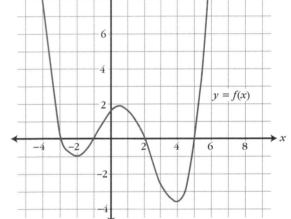

4. a. Find the maximum point for the part of the function between $x = -3$ and $x = 3$. That is, what point with an $x$-coordinate between $-3$ and 3 has the largest $y$-coordinate?

    b. Find the minimum point for the part of the function between $x = -3$ and $x = 3$.

5. Solve the inequality $f(x) > 0$. That is, find the values for $x$ that give a positive value for $y$. Describe all possible answers.

# Homework 29

# Functioning in the Math World

## Part I: Expressions, Graphs, Tables, and Situations

Algebraic expressions, graphs, and In-Out tables are three ways of representing the concept that mathematicians refer to using the word **function.** Mathematicians often blur the distinctions among these three representations, referring to them all as "the function."

Functions also are often connected to real-life situations. In this assignment, you will look at how these four ideas—expressions, graphs, tables, and situations—are related to one another. (Sometimes people use the phrase *rule of four* to refer to these four ways of thinking about functions.)

1. The area of a square is determined by the length of any of its sides. For instance, if the length of a side is 7 inches, then the area is 49 square inches. Therefore, we can say that the area is *a function of* this length.

   a. Make an In-Out table to go with this situation and fill in several rows for this table. (The values 7 and 49 would make up one row of your table.)

*Continued on next page*

Interactive Mathematics Program                                           85

b. Express the relationship between area and length in terms of an equation, explaining any variables you use.

c. Make a graph of your equation.

2. The equation $y = 3x + 1$ can be used to define a function.

a. Make an In-Out table to go with this equation and fill in several rows for this table.

b. Sketch the graph of the equation.

c. Create a situation for which this equation might be appropriate. Explain the role of the variables $x$ and $y$ in your situation.

## Part II: Another Mystery Graph

3. Study the graph of the function $h$ shown below and then answer the questions. (You should consider only the part of the function shown in this graph.)

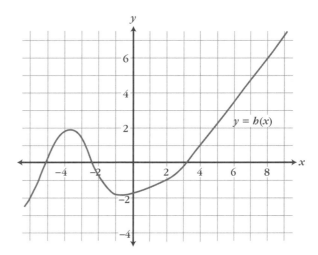

a. Estimate the value of $h(5)$.

b. Estimate the $y$-intercept of the function $h$.

*Continued on next page*

c. Estimate the *x*-intercepts of the function *h*.

d. Estimate all solutions to the equation $h(x) = -1$.

## Part III: Making Graphs That Fit Conditions

In these problems, you are asked to sketch graphs that fit certain conditions. Many graphs will work, but you need to create only one of them. *Note:* Most graphs are infinite, so the graph you actually draw will probably be only part of the total graph.

4. Sketch the graph for a function *f* that satisfies both of these conditions.

   • The function *f* has exactly three *x*-intercepts.

   • $f(2) = 5$

5. Sketch the graph for a function *k* that satisfies all of these conditions.

   • The function *k* has no *x*-intercepts.

   • $k(-1) = 2$

   • $k(3) = 1$

   • $k(5) = 6$

DAY 30

# *A Graphing Calculator Approach*

*Students work with graphing calculators to answer questions about equations and graphs.*

## Mathematical Topics

- Connecting graphs, tables, equations, and situations
- Interpreting graphs of functions
- Creating graphs that fit given conditions
- Solving equations using a graphing calculator

## Outline of the Day

### In Class

1. Discuss *Homework 29: Functioning in the Math World*
2. *The Graphing Calculator Solver*
   - Students use a graphing calculator to solve equations
   - No whole-class discussion of this activity is needed

### At Home

*Homework 30: A Solving Sampler*

## 1. Discussion of *Homework 29: Functioning in the Math World*

Questions 1 and 2 may be fairly routine for students, so use your judgment about whether to spend time discussing them. Question 3 is similar to yesterday's activity, *Mystery Graph,* so students' work on that activity should give you an idea of how much time to spend on this question.

Questions 4 and 5 involve a new kind of task so you should be sure to discuss them. Have one or two heart card students present their work on

Question 4, and then ask if other students got anything significantly different. Although the graphs may vary in detail, they should generally be similar. You can use a similar approach for Question 5.

The graphs shown here give a possible solution for each of these two questions.

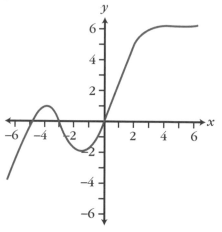 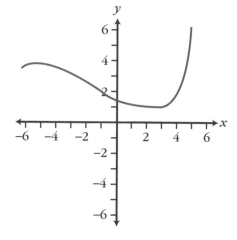

**Possible Solution to Question 4**    **Possible solution to Question 5**

*Note:* The supplemental problem *A Function—Not!* discusses the distinction between a function and an arbitrary set of number pairs. You may want to assign that problem as a follow-up to the homework.

## 2. The Graphing Calculator Solver

*The Graphing Calculator Solver* concludes the final segment of the unit, which focuses on the use of graphs to solve equations.

In recent activities, students began with functions defined by real-world problem situations and then developed and solved equations based on the functions. In this activity, students are confronted with equations without problem contexts.

If groups need a hint to get started on Question 1, you can ask what function they might graph to help with the equation. As a further hint, have them look back at their work on the activities *Where's Speedy?* and *To the Rescue.*

Question 4 is more difficult, because the expression on the right is not just a single number. As suggested in the hint, probably the most natural approach is to graph the functions defined by the two expressions $x^3 + 4x$ and $2x^2 + 7x - 1$ and then to look for the places where the graphs meet. Alternatively, students might replace the equation by an equivalent one, such as the equation obtained by moving all terms to one side of the equation, but you need not mention this if students don't bring it up.

No whole-class discussion of this activity is needed.

Here are the solutions to these equations.

- Question 1: $x = -4.1$ and $x = 1.6$

- Question 2: $x = -4.6$, $x = -1.3$, and $x = 1.9$

- Question 3: $x = -1.4$ and $x = 0.8$

- Question 4: $x = -1.2$, $x = 0.3$, and $x = 2.9$

You may want to tell students that in the Year 4 unit *The World of Functions,* they will study the general behavior of the graphs of certain families of functions. That knowledge will allow them to get all the solutions to many equations without the kind of clue provided here.

## Homework 30: A Solving Sampler

This overview assignment will form another part of students' portfolios for the unit.

**Kate Toomey carefully records the results of her partner's exploration of equations.**

# The Graphing Calculator Solver

In *Where's Speedy?* the expression $0.1t^2 + 3t$ told you how far Speedy would run in the first $t$ seconds. The expression was used to define the function $m(t)$.

You then wrote and solved equations with this expression to answer questions about Speedy. For example, you solved the equation $0.1t^2 + 3t = 200$ to find out how long it would take Speedy to run the first 200 meters of her part of the race. You may have used the graph of the function $m(t)$ to solve such equations.

*Continued on next page*

# Equations, Functions, and Graphs

You can use this method to solve equations even when the equation doesn't come from a real-life situation. For example, even if you saw the equation $0.1t^2 + 3t = 200$ out of context, you could still enter the expression $0.1X^2 + 3X$ into a graphing calculator to define a function and get a graph. You could then solve the equation (approximately) by using the trace feature to find the values of $X$ that make $Y$ equal to 200. (In addition to the solution you found earlier, there is also a negative solution for $X$, which you would ignore if $X$ represented time.)

# A Graph Is Like an Answer Key

There's nothing magical about the expression $0.1t^2 + 3t$. Any meaningful expression that can be entered into a graphing calculator can be used to define a function. Once you've entered the expression, the graph of the function becomes an answer key for an entire family of equations involving that expression. For example, the graph of Speedy's function doesn't just help you solve the equation $0.1t^2 + 3t = 200$. It also helps you solve $0.1t^2 + 3t = 100$, $0.1t^2 + 3t = 63$, or any similar equation.

# Your Task

Your task in this activity is to use the graphing calculator to solve the equations given below. Give your answers to the nearest tenth.

*Note:* For simplicity, these equations have been chosen so that their solutions all lie between $x = -5$ and $x = 5$.

1. $2x^2 + 5x + 7 = 20$

2. $x^3 + 4x^2 - 5x + 1 = 12$

3. $x^4 - x^3 + 3x^2 + 5x = 6$

4. $x^3 + 4x = 2x^2 + 7x - 1$ (*Hint:* Use the expressions on each side of the equation to define two different functions. Then think of the original equation as asking for the value of $x$ that gives the same function value for both functions.)

# Homework 30                    A Solving Sampler

You have used several ideas and methods for solving equations as part of this unit.

- Guess and check

- The mystery-bags model

- "Unscrambling" equations (equivalent equations)

- Graphing

In this activity, you will examine each of these methods as part of the preparation for your *Solve It!* portfolio.

1. Begin with the guess-and-check method.

    a. Summarize how the method works.

    b. Select an activity from the unit in which that method played an important role and attach that activity.

    c. Make up an equation for which you would use that method.

2. Do parts a, b, and c of Question 1 for the mystery-bags model.

3. Do parts a, b, and c of Question 1 for the equivalent-equations method.

4. Do parts a, b, and c of Question 1 for the graphing method.

# DAY 31 *Portfolios*

Students compile unit portfolios.

## Mathematical Topics

• Reviewing the unit and preparing portfolios

## Outline of the Day

### In Class

1. Discuss *Homework 30: A Solving Sampler*

2. *"Solve It!" Portfolio*

   • Students write cover letters and assemble portfolios for the unit

3. Remind students that unit assessments will take place tomorrow and tomorrow night

### At Home

Students prepare for end-of-unit assessment

## 1. Discussion of *Homework 30: A Solving Sampler*

You might have four volunteers present ideas, each discussing one of the four methods. Students can hold onto this assignment for inclusion in their portfolios.

## 2. *"Solve It!" Portfolio*

Tell students to read over the instructions in *"Solve It!" Portfolio* carefully and then take out all of their work from the unit. They will have done part of the selection process in *Homework 30: A Solving Sampler,* so their main task today is to write their cover letters.

If students do not complete their portfolios in class, you may want them to take the materials home and finish compiling the portfolios as homework. Be sure that they bring back the portfolio tomorrow with the cover letter as the first item. They should also bring any other work that they think will be of help on tomorrow's unit assessments. The remainder of their work can be kept at home.

## *Homework: Prepare for Assessments*

Students' homework for tonight is to prepare for tomorrow's assessments by reviewing the ideas of the unit.

*The graphing calculator helps Jason Mahoney and Caitlin Desmond in their work with nonlinear equations.*

# "Solve It!" Portfolio

Now that *Solve It!* is completed, it is time to put together your portfolio for the unit. This activity has three parts.

- Writing a cover letter summarizing the unit

- Choosing papers to include from your work in this unit

- Comparing this unit to Year 1 IMP units and to traditional algebra

## Cover Letter for "Solve It!"

Look back over *Solve It!* and describe the main mathematical ideas of the unit. This description should give an overview of how the key ideas were developed. In compiling your portfolio, you will be selecting some activities that you think were important in developing the key ideas of this unit. Your cover letter should include an explanation of why you selected particular items.

*Continued on next page*

# Papers from "Solve It!"

Your portfolio for *Solve It!* should contain these items.

- *Homework 30: A Solving Sampler.*
  Include both what you wrote about the different methods for solving equations and the sample activities you chose.

- *Get It Straight.*
  Include your write-up of your work on this activity.

- *Homework 24: A Distributive Summary.*

- A Problem of the Week.
  Select one of the three POWs you completed during this unit (*A Digital Proof* or *Tying the Knots* or *Divisor Counting*).

- Other high-quality work.
  Select one or two other pieces of work that represent your best efforts. (These can be any work from the unit—Problem of the Week, homework, classwork, presentation, and so on.)

# "Solve It!" and Algebra

This unit is more traditional than most of the IMP units. For example, it doesn't have a central problem, and it involves manipulating algebra symbols. Discuss your reaction to this type of unit. You might comment on these issues.

- How did you like this unit compared to Year 1 units with a central problem?

- Are you glad you did a unit emphasizing these traditional skills? Why or why not?

- How does the material in this unit compare with your idea of what algebra is?

# Final Assessments

*Students do the in-class assessment and can begin the take-home assessment.*

## Special Materials Needed

- *In-Class Assessment for "Solve It!"*
- *Take-Home Assessment for "Solve It!"*

## Outline of the Day

### In Class

Introduce assessments

- Students do *In-Class Assessment for "Solve It!"*
- Students begin *Take-Home Assessment for "Solve It!"*

### At Home

Students complete *Take-Home Assessment for "Solve It!"*

## End-of-Unit Assessments

*Note:* The in-class assessment is intentionally quite short so that time pressures will not affect students' performance. The IMP *Teaching Handbook* contains general information about the purpose of the end-of-unit assessments and ways to use them.

Tell students that today they will get two tests—one that they will finish in class and one that they can start in class and will be able to finish at home. The take-home part should be handed in tomorrow.

Tell students that they are allowed to use graphing calculators, notes from previous work, and so forth, when they do the assessments. (They will have to do without graphing calculators when they complete the take-home portion at home unless they have their own.)

These assessments are provided separately in Appendix B for you to duplicate.

# In-Class Assessment for "Solve It!"

1. Solve each of these equations using the idea of equivalent equations. Show and explain each step of your work.

   a. $22M + 19 = 13M + 41$

   b. $7(x - 3) + 29 = 3(x + 2) + 7$

   c. $4y - (3y - 6) = 2(y + 4)$

2. The formula $V = x(12 - 2x)^2$ gives the volume for a box built by cutting square corners of side $x$ from a 12-inch-by-12-inch sheet of cardboard.

   a. Based on this formula, write an equation that you could use to find out how big to make the corners if you wanted the volume to be 120 cubic inches.

   b. Solve your equation and explain your method. Give all solutions to the equation to the nearest tenth of an inch. *Note:* If you use a graphing calculator, be sure to choose settings for the viewing window that are appropriate to the situation.

# Take-Home Assessment for *"Solve It!"*

## Part I: Dark and Stormy Shadows

On a dark and stormy night, you are standing near a 20-foot-tall lamppost. You figure that if your shadow from the lamppost is really long, you won't be quite as fearful of the storm. Assume that you are $5\frac{1}{2}$ feet tall. You want to know how far you should stand from the lamppost in order to cast an 8-foot shadow.

   1. Write an equation whose solution will solve this problem. Explain why your equation represents the problem.

   2. Solve the equation. Explain what your answer means in terms of the problem.

## Part II: Mystery Graph

This graph shows the variable y as a function of x, but it doesn't give a formula for this function. Instead of being described by a formula, the graph is labeled with the generic function equation $y = g(x)$. (You should consider only the part of the function shown in this graph.)

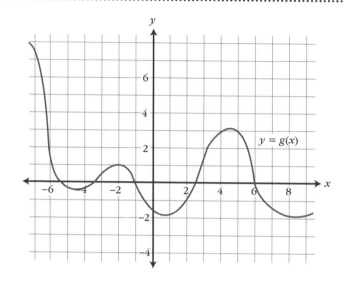

Answer these questions based on the graph, giving approximate answers if necessary.

3. a. What is the value of $g(-2)$?

    b. Describe the process by which you found the answer to Question 1a.

4. a. Give all possible solutions to the equation $g(x) = 2$.

    b. Describe the process by which you found the answer to Question 2a.

5. Give all $x$-intercepts for the function $g$.

6. Give all $y$-intercepts for the function $g$.

## Part III: Make It Up!

For each of these equations, make up a problem that it could represent.

7. $5x = 30$

8. $3(x - 4) = 105$

# Summing Up

## Mathematical Topics

• Summarizing the unit

## Outline of the Day

1. Discuss unit assessments
2. Sum up the unit

*Note:* The assessment discussions and unit summary are presented as if they take place on the day following the assessments, but you may prefer to delay this material until after you have looked over students' work on the assessments. These discussion ideas are included here to remind you that some time should be allotted for this type of discussion.

## 1. Discussion of Unit Assessments

You can have students volunteer to explain their work on each of the problems. Encourage questions and alternate explanations from other students.

### • In-class assessment

On Question 1, students should show each step of the manipulations. Although the problem doesn't explicitly ask students to check their answer, you should have someone do this in the discussion.

On Question 2a, students should have an explicit equation, probably $x(12 - 2x)^2 = 120$.

On Question 2b, they should give a clear explanation for how they solved this equation. A guess-and-check approach is certainly acceptable. If students used a graphing calculator, they should explain how they used it. For example, they might say something like

> "I graphed $y = x(12 - 2x)^2$. I adjusted the screen to go up to $y = 120$ and saw that there seemed to be three points with a $y$-coordinate close to 120. I zoomed in to get the $x$-coordinates more closely, and got 1.4, 2.6, and 7.9."

If students found only the first two of these points, they may have reasoned that they should only consider *x*-values between 0 and 6, which is a sensible approach. Because this unit has not tried to teach students about the general shape of cubic functions, there is no reason for students to expect the graph to go back up, and they should probably not be penalized for getting only the first two solutions to the equation.

• *Take-home assessment*

Students will probably use a diagram on Question 1 of Part I and come up with an equation equivalent to $\frac{8}{8+x} = \frac{5.5}{20}$. They should find that the person should stand about 21 feet from the lamppost.

On Part II (Questions 3 through 6), the main issues are students' ability to interpret and use the graph and to work with function notation.

On Part III (Questions 7 and 8), you will probably get a wide variety of responses. Be sure that students give the meaning of their variables in the problem situations.

# 2. Unit Summary

You can have students share their portfolio cover letters, as well as their work on *Homework 30: A Solving Sampler* and *Homework 24: A Distributive Summary*, as a way to start a summary discussion of the unit. You can also let students share their ideas about *Solve It!* and algebra from the last part of their portfolios.

# Appendix A

# Supplemental Problems

This appendix contains a variety of additional activities that can be used to supplement the regular unit material. These activities fall roughly into two categories.

- Reinforcements, which are intended to increase students' understanding of and comfort with concepts, techniques, and methods that are discussed in class and that are central to the unit

- Extensions, which allow students to explore ideas beyond the basic unit and which sometimes deal with generalizations or abstractions of ideas that are part of the main unit

The supplemental activities are given here and in the student materials in the approximate sequence in which you might use them in the unit. In the student book, they are placed following the regular materials for the unit. The discussion here makes specific recommendations about how each activity might work within the unit. For more ideas about the use of supplemental activities in the IMP curriculum, see the IMP *Teaching Handbook*.

- *What to Expect* and *Carlos and Betty* (reinforcements)

    These problems offer further opportunities for students to review ideas from *The Game of Pig*. They can be used anytime after the Day 3 discussion of *Memories of Yesteryear*.

- *Ten Missing Digits* (extension)

    This is a more advanced version of *Is It a Digit?* and can be assigned after the Day 11 discussion of *POW 1: A Digital Proof*.

- *Same Expectations* (extension)

    In this activity, students apply the distributive property to understand why any number of trials can be used to compute expected value. The activity can be used after Day 15.

- *Preserve the Distributive Property* (extension)

    This activity offers an abstract perspective on the rule for multiplying two negative numbers, based on the distributive property, and might be assigned anytime after about Day 16.

- *The Locker Problem* (extension)

    This problem involves mathematical ideas similar to those from *POW 3: Divisor Counting.* You may want to use this as a follow-up or an alternative to that POW.

- *Who's Got an Equivalent?* and *Make It Simple* (reinforcements)

    These activities provide further practice with removing parentheses and simplifying algebraic expressions. They can be used after the Day 20 discussion of *Taking Some Out, Part II.*

- *Linear in a Variable* (extension)

    In this activity, students will extend their perspective on linearity, looking at equations that are not linear, but that are linear when considered in terms of a particular variable. *Linear in a Variable* is a good follow-up to *Homework 25: All by Itself.*

- *The Shadow Equation Revisited* (extension)

    In the Year 1 unit *Shadows,* students used an alternate pair of similar triangles to get around the complications of the equation that arose from their shadow diagram. In this activity, they are asked to work directly with the original equation. This activity might be assigned after *Homework 26: More Variable Solutions.*

- *A Function—Not!* (extension)

    This discussion of the formal difference between a function and an arbitrary set of pairs might work well following *Homework 29: Functioning in the Math World.*

*Solve It!*

# Supplemental Problems

*This page in the student book introduces the supplemental problems.*

As in Year 1, the supplemental problems for each unit pursue some of the themes and ideas that are important in that unit. Some of the supplemental problems in *Solve It!* continue the theme of looking back at ideas from Year 1. Others pursue ideas about equivalent expressions or follow up on ideas from the POWs. Here are some examples.

- *What to Expect* and *Carlos and Betty* give you more opportunities to work with the concept of expected value from the Year 1 unit *The Game of Pig*.

- *Same Expectations* and *Preserve the Distributive Property* give you a chance to extend your understanding of the distributive property and how it is used.

- *Ten Missing Digits* and *The Locker Problem* continue themes from *POW 1: A Digital Proof* and *POW 3: Divisor Counting*.

# What to Expect?

One of the problems from *Memories of Yesteryear* involved Al and Betty and the spinner shown below. This problem poses some more questions about that spinner.

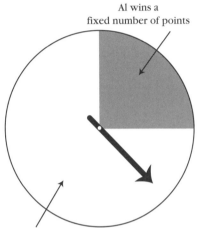

Al wins a
fixed number of points

Betty wins 2 points

1. a. If Al wins 4 points when the spinner lands on the gray area, what is his expected gain or loss per spin in the long run?

   b. If Al wins 10 points when the spinner lands on gray, what is his expected gain or loss per spin in the long run?

2. Al likes playing spinner games. He's willing to lose an average of $\frac{1}{4}$ point per spin in order to keep playing. Assume that Betty still wins 2 points when the spinner lands on white. What payoff should Al be willing to take each time the spinner lands on gray so that his expected loss is $\frac{1}{4}$ point per spin?

3. Make an In-Out table, based on the spinner shown above, in which the *In* is the amount Al wins when the spinner lands on gray and the *Out* is Al's average gain or loss per spin for that payoff. (Betty always gets 2 points when she wins.)

4. Find a rule for your In-Out table.

# Carlos and Betty

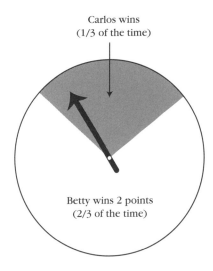

Carlos wins
(1/3 of the time)

Betty wins 2 points
(2/3 of the time)

Carlos also likes spinner games. He has a spinner like the one at the right, in which Betty wins when the spinner lands in the white area ($\frac{2}{3}$ of the time). She gets 2 points from Carlos every time she wins.

1. How many points should Carlos win when the spinner lands in the gray area in order for the game to be fair? Explain how you got your answer.

2. Betty thinks it's fun to play spinner games. In fact, she is willing to lose points in the long run in order to keep Carlos interested in playing.

   How many points should she give him for the spins he wins if she wants his average gain in the long run to be

   a. $\frac{1}{10}$ point per spin?

   b. $\frac{1}{2}$ point per spin?

   Explain how you got your answers.

*Hint on both Question 1 and Question 2:* Experiment with various payoff amounts, find Carlos' expected value for each, and make an In-Out chart.

# *Ten Missing Digits*

In *Is It a Digit?* you had to fill in five empty boxes, labeled 0 through 4, in a way that satisfied certain conditions. In this problem, you have to solve a harder version of that problem. Specifically, consider the ten boxes below:

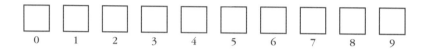

The rules are that you must put a digit from 0 to 9 in each of the boxes so that these conditions hold.

- The digit you put in the box labeled "0" must be the same as the number of 0's you use.

- The digit you put in the box labeled "1" must be the same as the number of 1's you use.

- The digit you put in the box labeled "2" must be the same as the number of 2's you use, and so on.

As in *Is It a Digit?* you are allowed to use the same digit more than once.

There may be more than one solution to this problem, so part of your task is to show that you have all the possible answers.

# Same Expectations

You may have noticed in your work with expected value in *The Game of Pig* that it didn't matter how many games or how many spins you used as "the long run." Here's your chance to see why.

The spinner shown here is the same as in Question 3 of *Memories of Yesteryear,* except that now it shows Al winning 5 points each time the spinner lands on gray. As before, Betty wins 2 points each time the spinner lands on white.

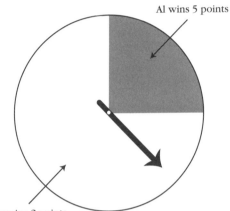

Al wins 5 points

Betty wins 2 points

1. Suppose Al and Betty do a total of 100 spins. Assume that the results follow the probabilities perfectly and find Betty's average gain per spin. Your expression should show your reasoning in terms of the 100 spins.

2. Repeat Question 1 but use 1000 spins this time.

3. Repeat Question 1 but this time use the variable $N$ for the number of spins. Use a little algebra to prove that if the results follow the probabilities perfectly, then Betty's average gain per spin is the same no matter how many spins there are.

# Preserve the Distributive Property

Some of the rules for multiplication with integers make intuitive sense, but others can be confusing. For example, the fact that $3(-2) = -6$ can be explained in terms of repeated addition:

$$3(-2) = -2 + (-2) + (-2)$$

The fact that $-2 + (-2) + (-2) = -6$ also seems reasonable, so we have $3(-2) = -6$.

It also seems intuitively reasonable to most people that $0(-2) = 0$. But many people have a hard time understanding why the product of two negative numbers should be positive. It turns out that the distributive property can provide an explanation for this fact.

Your first task in this activity is to use the expression $(-3 + 3)(-2)$ to show that $(-3)(-2) = 6$. The key idea is to evaluate $(-3 + 3)(-2)$ in two different ways, using the distributive property in one of the ways. Then generalize this example to explain why the product of any two negative numbers must be positive. You may assume that the product of a positive number and a negative number is negative.

# The Locker Problem

Louise is walking through the hallway of her school past the row of lockers on the first day of school. The lockers are numbered from 1 to 100. When Louise gets to the lockers, they are all open. Absentmindedly, Louise closes all the even-numbered lockers—the multiples of 2—as she walks by.

A few minutes later, Jeremy comes by. He decides to touch only those lockers whose numbers are multiples of 3. If one of these lockers is open when he goes by, he closes it, and if it's closed, he opens it. (For example, Louise left locker 3 open, so Jeremy closes it. Louise closed locker 6, so Jeremy opens it, and so on.)

Then another student comes by, and this student changes the doors on all the lockers whose numbers are multiples of 4. Then another student changes the doors on lockers whose numbers are multiples of 5, and so on, until finally a student comes by who changes only locker 100.

The question is,

*Which lockers are open at the end of the process?*

You should not only determine which lockers end up open but also find an explanation for the result. Once you're done, explain which lockers would end up open if the locker numbers went up to 1000. (Assume that the last student changes only locker 1000.)

Interactive Mathematics Program 99

# Who's Got an Equivalent?

For each of the expressions below:

- Find an expression without parentheses that is equivalent to the given one.

- Explain why the two expressions are equivalent. (Thinking about the hot-and-cold-cube model may help.)

1. $12 - (a + 7)$

2. $26 - (12 - 3t)$

3. $41 - 2(b + 1)$

# Make It Simple

The task of removing parentheses from an expression being subtracted is a tricky one, and not easy to explain. When the expression in parentheses itself involves subtraction (as in Question 2 of *Who's Got an Equivalent?*), it's even harder.

1  Describe and explain the steps involved in simplifying these expressions.

    a.  $20 - 5(x + 3)$

    b.  $20 - 5(x - 3)$

2. The next expression can be simplified all the way to $4x + 7$. Show the process of simplifying it to that point.

$$6(3x + 4) - 4(x - 2) - 5(2x + 5)$$

# Linear in a Variable

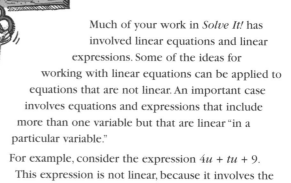

Much of your work in *Solve It!* has involved linear equations and linear expressions. Some of the ideas for working with linear equations can be applied to equations that are not linear. An important case involves equations and expressions that include more than one variable but that are linear "in a particular variable."

For example, consider the expression $4u + tu + 9$. This expression is not linear, because it involves the product of two variables, $t$ and $u$. But if $t$ were replaced by a specific number, the expression would become linear. For instance, substituting 3 for $t$ gives $4u + 3u + 9$, which is equivalent to the linear expression $7u + 9$.

The expression $4u + tu + 9$ is called **linear in $u$,** because replacing the other variable by a number gives a linear expression. You can use the distributive property to see that this expression is equivalent to $(4 + t)u + 9$. You can think of the sum $4 + t$ as the coefficient of $u$.

1. Solve the equation $4u + tu + 9 = 6t - 4$ for $u$ in terms of $t$.

2. Solve each of the equations below for the variable indicated. In each case, the equation is linear in that variable, even if the equation as a whole is not linear.

   a. Solve for $c$ in terms of $a$.

   $$ac + 3c = 5a - (c + 7)$$

   b. Solve for $w$ in terms of $u$ and $v$.

   $$4uw + v = 3w + v^2$$

# The Shadow Equation Revisited

Do you remember this picture from *Shadows* in Year 1? The small triangle and the big triangle in this diagram are similar, so the variables S, D, H, and L satisfy the equation

$$\frac{S}{S + D} = \frac{H}{L}$$

(Remember that S + D is the length of the horizontal side of the big triangle.)

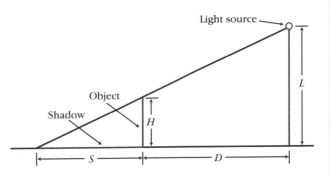

Your goal in *Shadows* was to express S directly as a function of L, H, and D. One way to accomplish this is to add a line segment to the diagram that creates another triangle similar to the original two. This gives another equation that is easier to use in solving for S.

1. Your first task in this activity is to use algebra instead of a new diagram to accomplish the *Shadows* goal.

   a. Find an equation equivalent to $\frac{S}{S+D} = \frac{H}{L}$ that expresses S in terms of L, H, and D. In other words, your equation should begin "S = " and have an expression involving the other three variables on the right side.
   (*Hint:* First find an equivalent equation that has no fractions in it, and then use the distributive property and factoring.)

   b. In *Lamppost Shadows*, Nelson was standing 20 feet from a 25-foot lamppost. Nelson is 6 feet tall. Use the equation you got in Question 1a to find the length of his shadow.

2. Algebra can be used to solve for other variables as well.

   a. Find an equation equivalent to $\frac{S}{S+D} = \frac{H}{L}$ that expresses D in terms of L, H, and S.

   b. Use your equation from Question 2a to find out where Nelson should stand in order to cast a 50-foot shadow.

# *A Function—Not!*

You have occasionally used In-Out tables in contexts in which these tables did not represent functions. For example, in *The Overland Trail,* a table showed the number of people in the group in one column and the amount used of a supply item in another. This relationship wasn't a function because groups of the same size might have used different amounts of the item.

The distinction between a function and an arbitrary set of pairs is sometimes important and might be stated like this:

> The term *function* is only applicable when the *Out* is completely determined by the *In.* In other words, for an In-Out table to represent a function, there can only be one value for the *Out* for any particular choice of the *In.*

## *The Vertical-Line Test*

One way to identify which graphs are graphs of functions is to apply the **vertical-line test:**

> For a graph to represent a function, no vertical line can go through more than one of its points.

*Note:* The definition of the term *function* permits many *In* numbers to have the same *Out,* so there is no "horizontal-line test" required for functions.

1. Identify which of these graphs are graphs of functions, and explain your decisions.

 a.

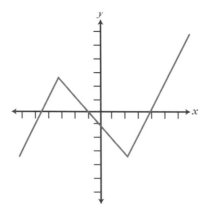

*Continued on next page*

b.

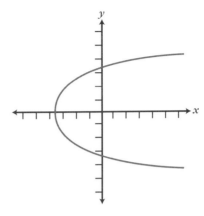

c.

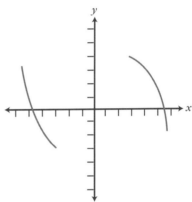

2. Explain why the vertical-line test tells you if a graph represents a function or not.

# Appendix B
# Blackline Masters

This appendix contains the following materials for use in the classroom:

- A large version of the graph used on Day 29 for introducing *Mystery Graph*. You can make an overhead transparency from this graph.

- The in-class and take-home unit assessments for *Solve It!* You will need to reproduce these for students for Day 32.

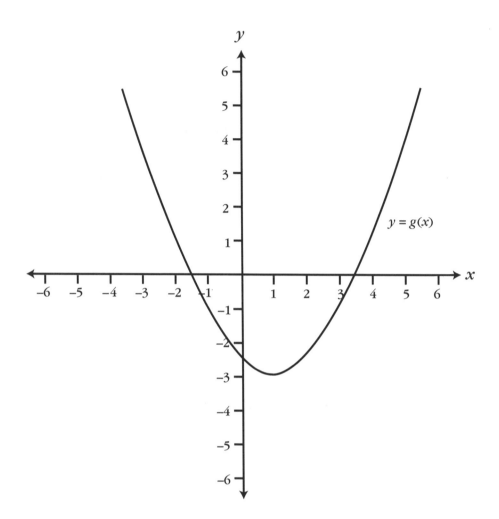

# In-Class Assessment for "Solve It!"

1. Solve each of these equations using the idea of equivalent equations. Show and explain each step of your work.

   a. $22M + 19 = 13M + 41$

   b. $7(x - 3) + 29 = 3(x + 2) + 7$

   c. $4y - (3y - 6) = 2(y + 4)$

2. The formula $V = x(12 - 2x)^2$ gives the volume for a box built by cutting square corners of side $x$ from a 12-inch-by-12-inch sheet of cardboard.

   a. Based on this formula, write an equation that you could use to find out how big the sides of the corners should be if you want the volume to be 120 cubic inches.

   b. Solve your equation and explain your method. Give all solutions to the equation to the nearest tenth of an inch. *Note:* If you use a graphing calculator, be sure to choose settings for the viewing window that are appropriate to the situation.

# Take-Home Assessment for *Solve It!*

## *Part I: Dark and Stormy Shadows*

On a dark and stormy night, you are standing near a 20-foot-tall lamppost. You figure that if your shadow from the lamppost is really long, you won't be quite as fearful of the storm. Assume that you are $5\frac{1}{2}$ feet tall. You want to know how far you should stand from the lamppost in order to cast an 8-foot shadow.

1. Write an equation whose solution will solve this problem. Explain why your equation represents the problem.

2. Solve the equation. Explain what your answer means in terms of the problem.

# Part II: Mystery Graph

This graph shows the variable *y* as a function of *x*, but it doesn't give a formula for this function. Instead of being described by a formula, the graph is labeled with the generic function equation $y = g(x)$. (You should consider only the part of the function shown in this graph.)

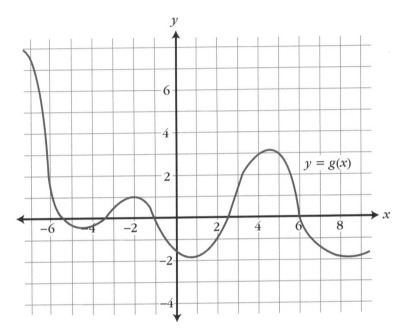

Answer these questions based on the graph, giving approximate answers if necessary.

3. a. What is the value of $g(-2)$?

    b. Describe the process by which you found the answer to Question 1a.

4. a. Give all possible solutions to the equation $g(x) = 2$.

    b. Describe the process by which you found the answer to Question 2a.

5. Give all *x*-intercepts for the function *g*.

6. Give all *y*-intercepts for the function *g*.

# Part III: Make It Up!

For each of these equations, make up a problem that it could represent.

7. $5x = 30$

8. $3(x - 4) = 105$

# Glossary

This is the glossary for all five units of IMP Year 2.

*Absolute growth*   The growth of a quantity, usually over time, found by subtracting the initial value from the final value. Used in distinction from **percentage growth.**

*Additive law of*
*exponents*   The mathematical principle which states that the equation

$$A^B \cdot A^C = A^{B+C}$$

holds true for all numbers *A*, *B*, and *C* (as long as the expressions are defined).

*Altitude of a*
*parallelogram*
*or trapezoid*   A line segment connecting two parallel sides of the figure and perpendicular to these two sides. Also, the length of such a line segment. Each of the two parallel sides is called a **base** of the figure.

Examples: Segment $\overline{KL}$ is an altitude of parallelogram *GHIJ*, with bases $\overline{GJ}$ and $\overline{HI}$ and segment $\overline{VW}$ is an altitude of trapezoid *RSTU*, with bases $\overline{RU}$ and $\overline{ST}$.

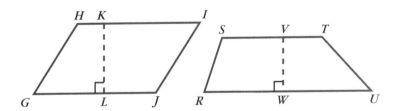

| | |
|---|---|
| *Altitude of a triangle* | A line segment from any of the three vertices of a triangle, perpendicular to the opposite side or to an extension of that side. Also, the length of such a line segment. The side to which the perpendicular segment is drawn is called the **base** of the triangle and is often placed horizontally. |

Example: Segment $\overline{AD}$ is an altitude of triangle *ABC*. Side *BC* is the base corresponding to this altitude.

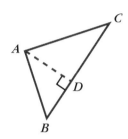

| | |
|---|---|
| *Base* | The side of a triangle, a parallelogram, or a trapezoid to which an altitude is drawn. For **base of a prism,** see *The World of Prisms* in the unit *Do Bees Build It Best?* |
| *Chi-square statistic* | A number used for evaluating the statistical significance of the difference between observed data and the data that would be expected under a specific hypothesis. The chi-square ($\chi^2$) statistic is defined as a sum of terms of the form |

$$\frac{(\text{observed} - \text{expected})^2}{\text{expected}}$$

with one term for each observed value.

| | |
|---|---|
| *Composite number* | A counting number having more than two whole-number divisors. |
| *Cosecant* | See *A Trigonometric Summary* in *Do Bees Build It Best?* |
| *Cosine* | See *A Trigonometric Summary* in *Do Bees Build It Best?* |
| *Cotangent* | See *A Trigonometric Summary* in *Do Bees Build It Best?* |
| *Dependent equations* | See **system of equations.** |

| | |
|---|---|
| *Distributive property* | The mathematical principle which states that the equation $a(b + c) = ab + ac$ holds true for all numbers $a$, $b$, and $c$. |
| *Edge* | See **polyhedron.** |
| *Equivalent equations (or inequalities)* | A pair of equations (or inequalities) that have the same set of solutions. |
| *Equivalent expressions* | Algebraic expressions that give the same numerical value no matter what values are substituted for the variables. Example: $3n + 6$ and $3(n + 2)$ are equivalent expressions. |
| *Expected number* | The value that would be expected for a particular data item if the situation perfectly fit the probabilities associated with a given hypothesis. |
| *Face* | See **polyhedron.** |
| *Factoring* | The process of writing a number or an algebraic expression as a product. Example: The expression $4x^2 + 12x$ can be factored as the product $4x(x + 3)$. |
| *Feasible region* | The region consisting of all points whose coordinates satisfy a given set of constraints. A point in this set is called a **feasible point.** |
| *Geometric sequence* | A sequence of numbers in which each term is a fixed multiple of the previous term. Example: The sequence $2, 6, 18, 54, \ldots$, in which each term is 3 times the previous term, is a geometric sequence. |
| *Hypothesis* | Informally, a theory about a situation or about how a certain set of data is behaving. Also, a set of assumptions used to analyze or understand a situation. |
| *Hypothesis testing* | The process of evaluating whether a hypothesis holds true for a given population. Hypothesis testing usually involves statistical analysis of data collected from a sample. |

| | |
|---|---|
| *Inconsistent equations* | See **system of equations.** |
| *Independent equations* | See **system of equations.** |
| *Inverse trigonometric function* | Any of six functions used to determine an angle if the value of a trigonometric function is known. |
| | Example: For $x$ between 0 and 1, the inverse sine of $x$ (written $\sin^{-1}x$) is defined to be the angle between 0° and 90° whose sine is $x$. |
| *Lateral edge or face* | See *The World of Prisms* in *Do Bees Build It Best?* |
| *Lateral surface area* | See *The World of Prisms* in *Do Bees Build It Best?* |
| *Law of repeated exponentiation* | The mathematical principle which states that the equation |

$$\left(A^B\right)^C = A^{BC}$$

holds true for all numbers $A$, $B$, and $C$ (as long as the expressions are defined).

| | |
|---|---|
| *Linear equation* | For two variables, an equation whose graph is a straight line. More generally, an equation stating that two linear expressions are equal. |
| *Linear expression* | For a single variable $x$, an expression of the form $ax + b$, where $a$ and $b$ are any two numbers, or any expression equivalent to an expression of this form. For more than one variable, any sum of linear expressions in those variables (or an expression equivalent to such a sum). |
| | Example: $4x - 5$ is a linear expression in one variable; $3a - 2b + 7$ is a linear expression in two variables. |
| *Linear function* | For functions of one variable, a function whose graph is a straight line. More generally, a function defined by a linear expression. |
| | Example: The function $g$ defined by the equation $g(t) = 5t + 3$ is a linear function in one variable. |

| | |
|---|---|
| *Linear inequality* | An inequality in which both sides of the relation are linear expressions. |
| | Example: The inequality $2x + 3y < 5y - x + 2$ is a linear inequality. |
| *Linear programming* | A problem-solving method that involves maximizing or minimizing a linear expression, subject to a set of constraints that are linear equations or inequalities. |
| *Logarithm* | The power to which a given base must be raised to obtain a given numerical value. |
| | Example: The expression $\log_2 28$ represents the solution to the equation $2^x = 28$. Here, "log" is short for *logarithm*, and the whole expression is read "log, base 2, of 28." |
| *Net* | A two-dimensional figure that can be folded to create a three-dimensional figure. |

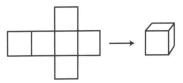

Example: The figure on the left is a net for the cube.

| | |
|---|---|
| *Normal distribution* | See *Normal Distribution and Standard Deviation Facts* in *Is There Really a Difference?* |
| *Null hypothesis* | A "neutral" assumption of the type that researchers often adopt before collecting data for a given situation. The null hypothesis often states that there are no differences between two populations with regard to a given characteristic. |
| *Order of magnitude* | An estimate of the size of a number based on the value of the exponent of 10 when the number is expressed in scientific notation. |
| | Example: The number 583 is of the second order of magnitude because it is written in scientific notation as $5.83 \cdot 10^2$, using 2 as the exponent for the base 10. |

| | |
|---|---|
| *Parallelogram* | A quadrilateral in which both pairs of opposite sides are parallel. |

Example: Polygons *ABCD* and *EFGH* are parallelograms.

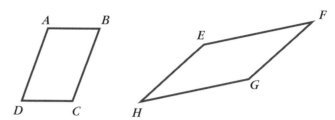

| | |
|---|---|
| *Percentage growth* | The proportional rate of increase of a quantity, usually over time, found by dividing the absolute growth in the quantity by the initial value of the quantity. Used in distinction from **absolute growth.** |

| | |
|---|---|
| *Polygon* | A closed two-dimensional figure consisting of three or more line segments. The line segments that form a polygon are called its sides. The endpoints of these segments are called **vertices** (singular: **vertex**). |

Examples: All the figures below are polygons.

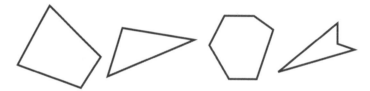

| | |
|---|---|
| *Polyhedron* | A three-dimensional figure bounded by intersecting planes. The polygonal regions formed by the intersecting planes are called the **faces** of the polyhedron, and the sides of these polygons are called the **edges** of the polyhedron. The points that are the vertices of the polygons are also **vertices** of the polyhedron. |

Example: The figure below shows a polyhedron. Polygon *ABFG* is one of its faces, segment $\overline{CD}$ is one of its edges, and point *E* is one of its vertices.

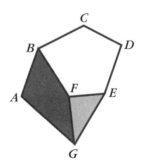

*Population*    A set (not necessarily of people) involved in a statistical study and from which a sample is drawn.

*Prime factorization*    The expression of a whole number as a product of prime factors. If exponents are used to indicate how often each prime is used, the result is called the **prime power factorization.**

Example: The prime factorization for 18 is $2 \cdot 3 \cdot 3$. The prime power factorization for 18 is $2^1 \cdot 3^2$.

*Prime number*    A counting number that has exactly two whole-number divisors, 1 and itself.

*Prism*    A type of polyhedron in which two of the faces are parallel and congruent. For details and related terminology, see *The World of Prisms* in *Do Bees Build It Best?*

*Profit line*    In the graph used for a linear programming problem, a line representing the number pairs that give a particular profit.

*Pythagorean theorem*    The principle for right triangles which states that the sum of the squares of the lengths of the two legs equals the square of length of the hypotenuse.

Example: In right triangle *ABC* with legs of lengths *a* and *b* and hypotenuse of length *c*, the Pythagorean theorem states that $a^2 + b^2 = c^2$.

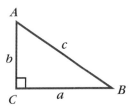

*Right rectangular prism*    See *The World of Prisms* in *Do Bees Build It Best?*

*Sample*    A selection taken from a population, often used to make conjectures about the entire population.

| | |
|---|---|
| *Sampling fluctuation* | Variations in data for different samples from a given population that occur as a natural part of the sampling process. |
| *Scientific notation* | A method of writing a number as the product of a number between 1 and 10 and a power of 10.<br><br>Example: The number 3158 is written in scientific notation as $3.158 \cdot 10^3$. |
| *Secant* | See *A Trigonometric Summary* in *Do Bees Build It Best?* |
| *Sine* | See *A Trigonometric Summary* in *Do Bees Build It Best?* |
| *Standard deviation* | See *Normal Distribution and Standard Deviation Facts* in *Is There Really a Difference?* |
| *Surface area* | The amount of area that the surfaces of a three-dimensional figure contain. |
| *System of equations* | A set of two or more equations being considered together. If the equations have no common solution, the system is **inconsistent.** Also, if one of the equations can be removed from the system without changing the set of common solutions, that equation is **dependent** on the others, and the system as a whole is also **dependent.** If no equation is dependent on the rest, the system is **independent.**<br><br>In the case of a system of two linear equations with two variables, the system is *inconsistent* if the graphs of the two equations are distinct parallel lines, *dependent* if the graphs are the same line, and *independent* if the graphs are lines that intersect in a single point. |
| *Tangent* | See *A Trigonometric Summary* in *Do Bees Build It Best?* |
| *Tessellation* | Often, a pattern of identical shapes that fit together without overlapping. |

| | |
|---|---|
| *Trapezoid* | A quadrilateral in which one pair of opposite sides is parallel and the other pair is not. |

Example: Polygons *KLMN* and *PQRS* are trapezoids.

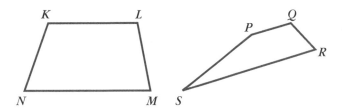

| | |
|---|---|
| *Trigonometry* | For a right triangle, the study of the relationships between the acute angles of the triangle and the lengths of the sides of the triangle. For details, see *A Trigonometric Summary* in the unit *Do Bees Build It Best?* |
| *Vertex* | See **polygon** and **polyhedron**. |
| *x-intercept* | A point where a graph crosses the *x*-axis. Sometimes, the *x*-coordinate of that point. |
| *y-intercept* | A point where a graph crosses the *y*-axis. Sometimes, the *y*-coordinate of that point. |

# Photographic Credits

## Teacher Book Classroom Photography

**7** Fresno High School, Dave Calhoun; **47** Michael Christensen, Pleasant Valley High School; **102** Don Cruser, Mendocino High School; **174** Fresno High School, Dave Calhoun; **214** Mike Bryant, Santa Maria High School; **221** David Harding, Lynne Alper, Milton High School; **226** David Harding, Milton High School

## Student Book Interior Photography

**3** Lincoln High School, Lori Green; **19** Santa Cruz High School, Kevin Drinkard, Lynne Alper; **38** Foothill High School, Cheryl Dozier; **67** Brookline High School, Priscilla Burbank-Schmitt, Carla Oblas; **72** Lumina Designworks, Terry Lockman; **77** Hillary Turner; **79** Fresno High School, Dave Calhoun

## Cover Photography and Cover Illustration

**Background** © Tony Stone Worldwide **Top left to bottom right** From *Alice in Wonderland* by Lewis Carroll; Hillary Turner; Hillary Turner; © Image Bank

## Front Cover Students

Colin Bjorklund, Liana Steinmetz, Sita Davis, Thea Singleton, Jenée Desmond, Jennifer Lynn Anker, Lidia Murillo, Keenzia Budd, Noel Sanchez, Seogwon Lee, Kolin Bonet (photographed by Hillary Turner)